The Impact of Chemical Sequencing on Filtration Performance

The mission of the Awwa Research Foundation (AwwaRF) is to advance the science of water to improve the quality of life. Funded primarily through annual subscription payments from over 1,000 utilities, consulting firms, and manufacturers in North America and abroad, AwwaRF sponsors research on all aspects of drinking water, including supply and resources, treatment, monitoring and analysis, distribution, management, and health effects.

From its headquarters in Denver, Colorado, the AwwaRF staff directs and supports the efforts of over 500 volunteers, who are the heart of the research program. These volunteers, serving on various boards and committees, use their expertise to select and monitor research studies to benefit the entire drinking water community.

Research findings are disseminated through a number of technology transfer activities, including research reports, conferences, videotape summaries, and periodicals.

The Impact of Chemical Sequencing on Filtration Performance

Prepared by:
Dean Gregory and **Kenneth Carlson, Ph.D.**
Colorado State University
Fort Collins, CO 80523-1372

and

William Bellamy, Ph.D.
CH2M HILL, Inc.
Englewood, CO 80112

Sponsored by:
Awwa Research Foundation
6666 West Quincy Avenue
Denver, CO 80235-3098

Published by the

Awwa Research Foundation
Advancing the Science of Water®

IWA Publishing

DISCLAIMER

This study was funded by the Awwa Research Foundation (AwwaRF). AwwaRF assumes no responsibility for the content of the research study reported in this publication or for the opinions or statements of fact expressed in the report. The mention of trade names for commercial products does not represent or imply the approval or endorsement of AwwaRF. This report is presented solely for informational purposes.

Published by IWA Publishing, Alliance House, 12 Caxton Street, London SW1H 0QS, UK

Telephone: +44 (0) 20 7654 5500; Fax: +44 (0) 20 7654 5555; Email: publications@iwap.co.uk;
Web: **www.iwapublishing.com**

AwwaRF report number 90945F. Originally published by AwwaRF for its subscribers in 2003.
IWA Publishing version published 2004.

ISBN 1 84339 863 x

CONTENTS

TABLE

FIGURES

FOREWORD

The Awwa Research Foundation is a nonprofit corporation that is dedicated to the implementation of a research effort to help utilities respond to regulatory requirements and traditional high-priority concerns of the industry. The research agenda is developed through a process of consultation with subscribers and drinking water professionals. Under the umbrella of a Strategic Research Plan, the Research Advisory Council prioritizes the suggested projects based upon current and future needs, applicability, and past work; the recommendations are forwarded to the Board of Trustees for final selection. The foundation also sponsors research projects through the unsolicited proposal process; the Collaborative Research, Research Applications, and Tailored Collaboration programs; and various joint research efforts with organizations such as the U.S. Environmental Protection Agency, the U.S. Bureau of Reclamation, and the Association of California Water Agencies.

This publication is a result of one of these sponsored studies, and it is hoped that its findings will be applied in communities throughout the world. The following report serves not only as a means of communicating the results of the water industry's centralized research program but also as a tool to enlist the further support of the nonmember utilities and individuals.

Projects are managed closely from their inception to the final report by the foundation's staff and large cadre of volunteers who willingly contribute their time and expertise. The foundation serves a planning and management function and awards contracts to other institutions such as water utilities, universities, and engineering firms. The funding for this research effort comes primarily from the Subscription Program, through which water utilities subscribe to the research program and make an annual payment proportionate to the volume of water they deliver and consultants and manufacturers subscribe based on their annual billings. The program offers a cost-effective and fair method for funding research in the public interest.

A broad spectrum of water supply issues is addressed by the foundation's research agenda: resources, treatment and operations, distribution and storage, water quality and analysis, toxicology, economics, and management. The ultimate purpose of the coordinated effort is to assist water suppliers to provide the highest possible quality of water economically and reliably. The true benefits are realized when the results are implemented at the utility level. The foundation's trustees are pleased to offer this publication as a contribution toward that end.

This report shows that the use of a higher coagulation pH may be particularly advantageous for utilities that encounter periods of rapidly changing water quality conditions such as runoff events. The use of relatively high coagulation pH conditions along with alum doses optimized by zeta potential appears to offer several advantages: 1) significantly improved particle removal in the settling process, 2) increased floc formation rates, 3) only slightly decreased TOC removal, compared to lower pH levels, and 4) a significantly wider operating range, in terms of zeta potential, for optimum filtration performance. The dual objectives of enhanced coagulation—optimization of NOM removal in addition to particles—can therefore be achieved at high pH as well as low pH.

Edmund G. Archuleta, P.E.
Chair, Board of Trustees
Awwa Research Foundation

James F. Manwaring, P.E.
Executive Director
Awwa Research Foundation

ACKNOWLEDGMENTS

The research team offers its sincere thanks to the PAC for their valuable suggestions, technical insight, and time devoted to this project. We also thank Traci Case of AwwaRF for her guidance and assistance in managing this project.

Special thanks are given to Kevin Gertig and Judith Billica of Fort Collins Utilities. Their participation and support in this research—through use of their pilot laboratory and analytical equipment—was greatly appreciated and allowed it to be successful. Thanks are also extended to the entire staff of the Fort Collins Water Treatment Facility, who were always helpful and willing to accommodate our experimental needs.

EXECUTIVE SUMMARY

INTRODUCTION

The ultimate purpose of the coagulation process in water treatment is to remove suspended particles and natural organic matter (NOM). From a public health and thus regulatory perspective, the suspended particles of greatest concern are pathogenic organisms such as the protozoa *Giardia* and *Cryptosporidium*. More recently, NOM removal has been emphasized to minimize the formation of disinfection by-products such as trihalomethanes (THMs) and haloacetic acids (HAAs), which can be formed upon the reaction between NOM and chlorine.

In raw waters lacking sufficient buffering capacity, particularly runoff from snowmelt or thunderstorms and those waters associated with reduced or low-pH environments, the use of alum requires alkalinity addition for pH control. Because runoff waters often contain relatively high concentrations of NOM, the alum demand in these waters is proportionally high, further emphasizing the need for pH control.

In treatment conditions that require alkalinity addition, the chemical doses and resulting coagulation pH levels are often determined by jar tests and settled water data. The target coagulation pH (for non-softening plants) is usually between 6.0 and 7.5. Within this range it is possible to minimize the concentration of soluble aluminum in finished water and effectively remove organic and particulate matter. However, within this pH range the overall effectiveness of the treatment process, particularly filtration performance, can also vary significantly. The sequence in which coagulation chemicals, including those used for acid/base adjustment, are added will determine coagulation pH and may significantly impact the filtration process.

RESEARCH OBJECTIVES/APPROACH

The primary objective of this research was to understand the impact of chemical sequencing and coagulation pH/zeta potential on filtration performance. The secondary goal was to determine effective coagulation processes—in terms of pH, zeta potential, and coagulant dose—for treating challenging (i.e., runoff-type) raw waters. In the first phase, bench-scale experiments were conducted to determine the impact of coagulation pH and alum dose on the sedimentation process and microfloc formation kinetics. To assess floc formation kinetics, an optical coagulation monitoring technique, photometric dispersion analysis, was used for a range of conditions and was evaluated as an indicator of overall process performance.

In the second phase of this project, pilot-scale filtration experiments were performed using 1) a river source during spring runoff, characterized by high turbidity, low alkalinity, and relatively high NOM levels, and 2) a typical reservoir surface water of relatively constant quality and containing moderate NOM levels. Filter effluent particle counts and total organic carbon (TOC) removal were used to determine the overall effectiveness of three different coagulation scenarios (pH = 6.0, 6.7, and 7.4 ± 0.1). The goal in a majority of the filtration experiments was to optimize alum dose at each pH level based on zeta potential (z.p.). However, runs were also conducted at non-optimum z.p. in an attempt to characterize the optimum operating range for each process. A z.p. of zero, indicative of destabilized particles, was used as the initial target. However, after several preliminary filtration runs were conducted at different z.p. values, it was clear that a positive z.p. was optimum for the experimental conditions being used. The z.p target was therefore adjusted to approximately +2.5 ± 1.5 mV.

In addition to single-stage alum addition processes, a two-stage alum coagulation process, which has been shown to enhance particle removal in previous research using runoff water (Carlson and Gregory, 2000) was also studied.

RESULTS AND DISCUSSION

Impact of pH/zeta potential on filtration performance

The initial bench-scale phase of this project demonstrated that higher coagulation pH and alum doses enhanced particle removal in the sedimentation process without significantly compromising TOC removal. In Figure ES.1, steady-state filter effluent particle counts from all the runoff and reservoir experiments are plotted against the corresponding coagulation zeta potentials from each run. Overall, the trends in filtration results showed that excellent filter effluent quality (total particles < 10/mL) was attainable for all three pH levels, but that higher pH may provide more consistent performance over a wider range of coagulation conditions. At low pH, steady-state particle counts increased rapidly as zeta potential moved from the optimum range (+2 to +4 mV) and towards zero. At high pH, optimum filtration performance was sustained over a wide range and was clearly less sensitive to zeta potential.

It should be noted that these filtration results are not due solely to solids mass loading rate, as steady-state particle counts equivalent to those attained at pH = 7.4 (settled turbidity ≈ 1 ntu) were often attained at pH = 6.0 and 6.7 (settled turbidity = 2 to 5 ntu) when the z.p. was optimized.

These results appear to indicate that the use of a higher coagulation pH allows process optimization over a broader range of zeta potential and alum doses. A higher concentration of adsorbed aluminum hydroxide on particle surfaces may promote a more uniform surface charge, thus decreasing overall repulsive forces and increasing the probability of particles attaching upon collision. Such a mechanism would likely enhance interparticle bridging and the development of floc suitable for solid-liquid separation processes. Although the low-pH filtration performance was not necessarily poor and may be acceptable, depending on site-specific water quality goals, the use of a higher coagulation pH may be particularly advantageous during periods of rapidly changing water quality conditions such as runoff events.

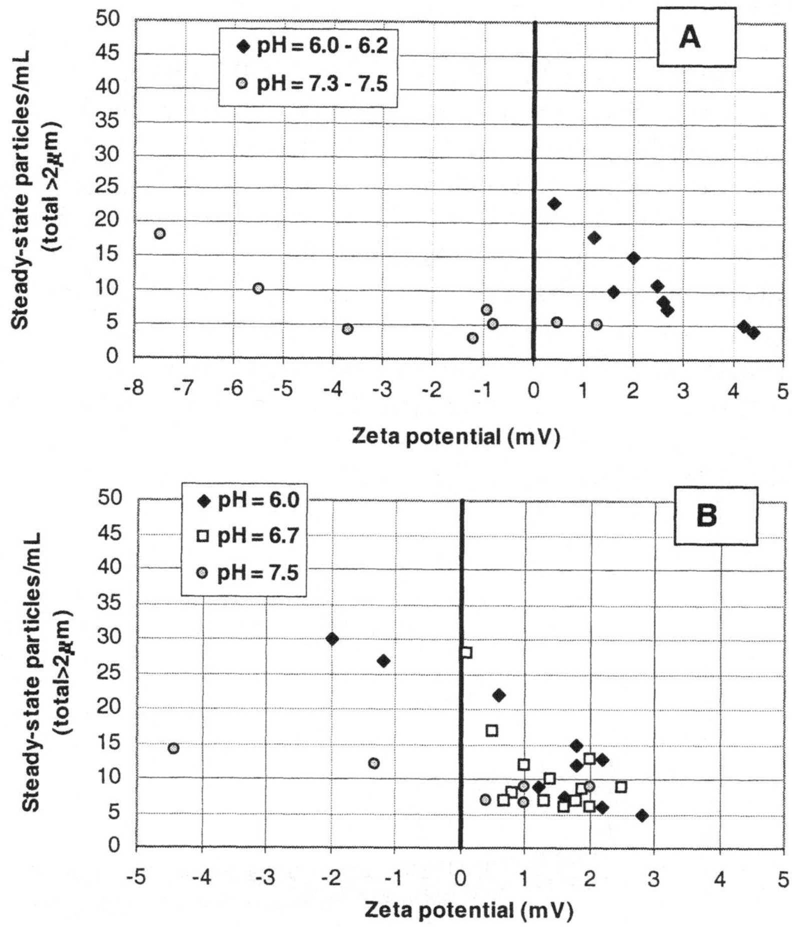

Figure ES.1 Steady-state filter effluent particle counts vs. coagulation z.p. for A) spring runoff water (TOC = 6 to 8 mg/L) and B) reservoir water (TOC = 3.5 mg/L)

Correlation of filtration results with photometric dispersion analyzer data

In general, it appeared that the flocculation index (FI) data correlated well with optimized filtration results and that use of a photometric dispersion analyzer (PDA) may be an alternative to jar testing for determining effective coagulation processes. In addition, settled water turbidity appeared to be directly related to the rate of floc formation, as indicated by the FI output from the PDA. If the rate of floc formation is indicative of the average suspended particle surface charge characteristics (i.e., the degree of destabilization achieved), then it seems reasonable to assume that FI would be related to filtration performance as well because floc particles become the primary particle receptors in filters after the initial ripening period (O'Melia, 1985). The results indicated that rate of FI increase was greatest for the alum dose that was optimum for filtration, in terms of zeta potential.

Two-stage alum addition processes

By use of a two-stage process, it was thought that the dual purposes of coagulation—removal of NOM and destabilization of suspended particles—could be accomplished more effectively. To optimize NOM removal, the target coagulation pH of the first stage was 6.0. To enhance particle removal, the pH was increased to 7.4 for the second stage. Although the research team was unable to apply this process during the runoff period, it was used during the reservoir water filtration experiments. As shown in Figure ES.2, the two-stage process was compared to the ambient process (pH = 6.7, no acid/base addition) and the high-pH process in head-to-head filtration runs. For the two-stage process, steady-state particle counts were approximately 20 percent lower than the ambient process and 50 percent lower than the high-pH process. Overall, the difference in steady-state particle counts between the two-stage and single-stage alum addition processes were not dramatic. However, use of the two-stage process did appear to improve filtration performance consistently.

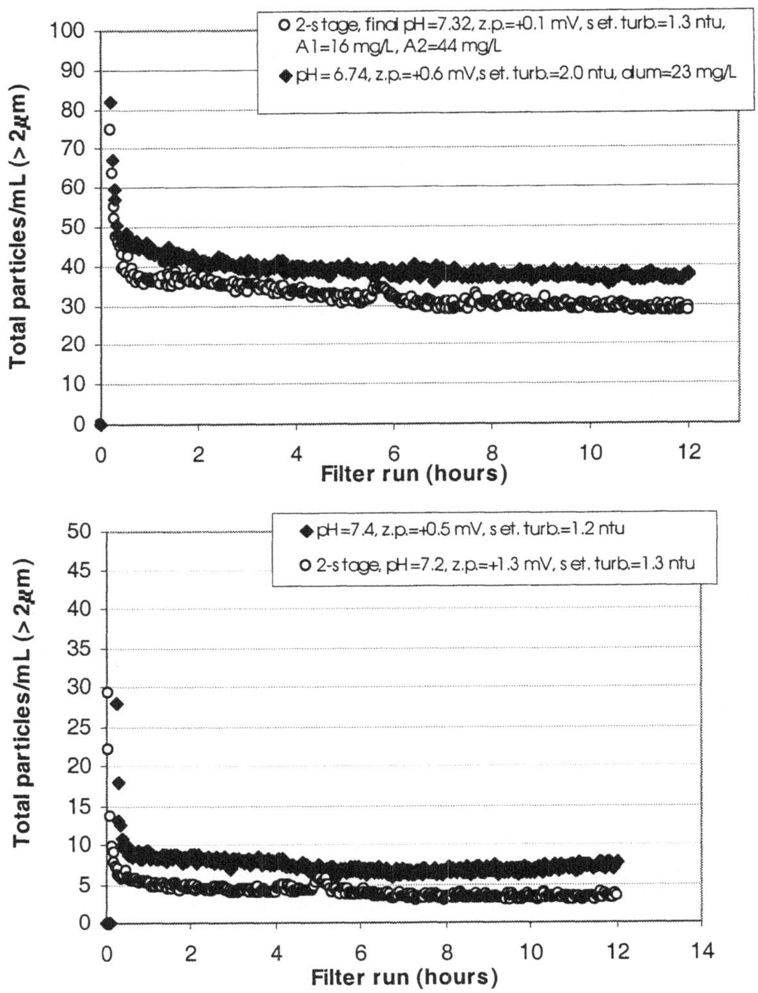

Figure ES.2 Comparison of filtration performance for two-stage and single-stage alum addition processes using reservoir water

CONCLUSIONS

Findings of the project included:

- In terms of overall process performance, the use of relatively high coagulation pH conditions along with alum doses optimized by zeta potential appeared to offer several advantages: 1) significantly improved particle removal in the settling process, 2) increased floc formation rates, 3) only slightly decreased TOC removal, compared to lower pH levels, and 4) a significantly wider operating range, in terms of zeta potential, for optimum filtration performance. The dual objectives of enhanced coagulation—optimization of NOM removal in addition to particles—can therefore be achieved at high pH as well as low pH.
- Although filtration performance in general was improved at higher pH levels for both water sources, the effect was greater for the more challenging runoff water. Because higher pH significantly increases the optimum zeta potential range for filtration, a practical implication of this result is that use of a high coagulation pH process may be the most effective short-term treatment strategy for rapidly changing, runoff-type waters containing low alkalinity and high NOM levels. Clearly, a high-pH process would not be practical, in terms of chemical feed costs and solids handling issues, for raw water that is effectively treated by conventional processes. If a high-pH process is unfeasible for treating runoff water, the zeta potential of coagulated water should be continuously monitored and the coagulant dose adjusted accordingly.
- The PDA appears to be a useful alternative to jar tests for determining optimum coagulant doses and quickly evaluating different coagulation scenarios. Overall, processes that the PDA indicated were optimum, in terms of the kinetics of floc development, were shown to be optimum in terms of filtration performance as well.
- Two-stage coagulation processes may enhance filtration performance for a wide range of raw water quality conditions. In this study, two-stage processes moderately improved filter effluent particle counts in reservoir water experiments. In previous research, two-stage alum addition was shown to be significantly more effective for runoff water.

CHAPTER 1
INTRODUCTION

The ultimate purpose of the coagulation process in water treatment is to remove suspended particles and natural organic matter (NOM). Suspended particles, commonly measured as turbidity, may include inorganic compounds such as clays as well as microbiological constituents such as algae and protozoa. NOM consists primarily of humic substances, including humic and fulvic acids, which result from the decomposition of terrestrial and aquatic biomass. From a public health and thus regulatory perspective, the suspended particles of greatest concern are pathogenic organisms such as the protozoa *Giardia* and *Cryptosporidium*. More recently, NOM removal has been emphasized to minimize the formation of disinfection by-products such as trihalomethanes (THMs) and haloacetic acids (HAAs), which can be formed upon the reaction between NOM and chlorine.

Of the coagulation chemicals most commonly used in North America—aluminum and iron salts—aluminum sulfate (alum) is the most common (AWWA, 1989). Alum ($Al_2[SO_4]_3 \cdot 14 H_2O$) is usually applied as the primary coagulant, often in conjunction with another coagulation chemical such as a cationic polymer which aids in building larger floc more suitable to solid-liquid separation processes. Due to its acidity, alum consumes naturally occurring alkalinity (generally HCO_3^-) and depresses the pH of water. In raw waters lacking sufficient buffering capacity, particularly runoff from snowmelt or thunderstorms and those waters associated with reduced or low-pH environments, the use of alum requires alkalinity addition for pH control. Because runoff waters often contain relatively high concentrations of NOM, the alum demand in these waters is proportionally high, further emphasizing the need for pH control.

In treatment conditions that require alkalinity addition, the chemical doses and resulting coagulation pH levels are often determined by jar tests and settled water data. The target coagulation pH (for non-softening plants) is usually between 6.0 and 7.5. Within this range it is possible to minimize the concentration of soluble aluminum in finished water and effectively remove organic and particulate matter, as demonstrated by Edzwald and Tobiason (1999). However, within this pH range the overall effectiveness of the treatment process, particularly filtration performance, can also vary significantly. The sequence in which coagulation chemicals, including those used for acid/base adjustment, are added will determine coagulation pH and may significantly impact the filtration process.

RESEARCH OBJECTIVES/APPROACH

The primary objective of this research was to understand the impact of chemical sequencing and coagulation pH/zeta potential on filtration performance. The secondary goal was to determine effective coagulation processes—in terms of pH, zeta potential (z.p.), and coagulant dose—for treating challenging (i.e., runoff-type) raw waters. In the first phase, bench-scale experiments were conducted to determine the impact of coagulation pH and alum dose on the sedimentation process and microfloc formation kinetics. To assess floc formation kinetics, an optical coagulation monitoring technique, photometric dispersion analysis, was used for a range of conditions and was evaluated as an indicator of overall process performance. In the second phase of this project, pilot-scale filtration experiments were performed using 1) a river source during spring runoff, characterized by high turbidity, low alkalinity, and relatively high NOM levels, and 2) a typical reservoir surface water of constant quality and containing moderate NOM levels. Filter effluent particle counts and total organic carbon (TOC) removal were used to determine the overall effectiveness of three different coagulation scenarios (pH = 6.0, 6.7, and 7.5).

In addition to single-stage alum addition processes, a two-stage alum coagulation process, which has been shown to enhance particle removal in previous research using runoff water (Carlson and Gregory, 2000) was also studied. However, a majority of the bench- and pilot-scale work has examined single-stage alum addition at different pH levels and related effects on microfloc formation rates, filtration performance, and NOM removal. The two-stage data included in this report are therefore limited. A two-stage process was applied during the filtration experiments in which reservoir water was used, but due to limited availability of runoff water, only single-stage processes were used during that period. During the reservoir water experiments, two pilot trains were utilized and the two-stage process was applied in parallel with an ambient, single-stage alum process (no acid/base addition).

CHAPTER 2
LITERATURE REVIEW

The subject of alum coagulation has been extensively researched and discussed in the literature. Numerous studies have documented 1) the role and relative concentration of various aluminum hydrolysis products in the pH range to be discussed here (6.0 to 7.5), and 2) the impact of dose, pH, and zeta potential on settled water quality. A majority of the filtration research has focused on developing theoretical models to predict particle removal. In general, these models have illustrated the relative importance of various physical and chemical filtration variables. Their correlation with laboratory filtration results, however, has been mixed.

The intent of this project was not to develop a theoretical model of filtration, but rather to understand some important chemical factors that influence coagulation and to extend these findings to filtration. It is hoped that the combination of laboratory- and pilot-scale results to be presented will further the knowledge of the impact of coagulation chemical sequencing on filtration. In the sections to follow, discussion will focus on previous studies that have: 1) clarified the role of different alum species in coagulation, particularly the importance of aluminum hydroxide sols ("sols" are defined as the sub-colloidal, poorly water-soluble, positively-charged aluminum hydroxide solids which form within 1 to 2 seconds following alum addition), 2) examined the effect of pH, coagulant dose, and zeta potential on sedimentation processes, 3) theoretically explored chemical and physical factors that influence filtration processes (primarily laboratory-scale studies), and 4) investigated the coagulation mechanisms that control NOM removal.

ALUMINUM SPECIES FORMED IN COAGULATION

The aluminum species formed upon the addition of alum to water are classified as monomers (e.g., Al^{3+}, $Al[OH]^{2+}$), polymers ($Al_2[OH]_2^{4+}$, $Al_{13}O_4[OH]_{24}^{7+}$), and the poorly-soluble precipitate ($Al(OH)_{3\ (S)}$) (Baes and Mesmer, 1976). The monomeric and polymeric species are thought to be relatively short-lived (<1 sec) intermediates formed during the stepwise hydrolysis of aluminum ion (Al^{3+}) (Hahn and Stumm, 1968; Letterman et al., 1973). The term *hydrolysis* describes the general reaction in which a proton is transferred from an acid to water. The equilibrium species formed by aluminum hydrolysis reactions are pH-dependent and are comprised of one or more of the following hydroxoaluminum complexes: $Al(OH)^{2+}$, $Al(OH)_2^{+}$, $Al(OH)_3$, and $Al(OH)_4^{-}$. As demonstrated previously (Hundt and O'Melia, 1988), in the pH and alum dose range of interest to this study, 6.0 to 7.5 and 15 to 100 mg/L, respectively, virtually all the aluminum is present as the solid precipitate $Al(OH)_{3\ (S)}$ at equilibrium. It was also shown in the same study that the presence of sulfate catalyzes the rapid formation of $Al(OH)_{3\ (S)}$ at pH > 4.6.

ROLE OF HIGHLY-CHARGED POLYMERIC SPECIES

In the pH range of 6.0 to 7.5, the two Al species that appear to control coagulation processes are the short-lived highly positively-charged polymeric intermediates and the poorly water soluble aluminum hydroxide sols that are formed within a few seconds after alum addition. Of these two species, there is substantial evidence that the positively-charged aluminum hydroxide sols, which themselves will aggregate into aluminum hydroxide microflocs after several seconds if not adsorbed to a solid surface, appear to be the most realistic species responsible for the destabilization of natural particles (Licskó, 1997; Dentel and Gossett, 1988). Although the existence of the highly-charged polynuclear complexes has been shown (Akitt and Farthing, 1978), their actual contribution to coagulation processes appears to be minor. In an attempt to determine the influence of these species, Klute (1990) performed bench-scale experiments, using a specialized mixing device, which demonstrated that the aluminum hydrolysis products formed within approximately 0.1 seconds after alum addition—presumably highly positively-charged polymers—were more effective at changing the surface charge of silica particles than the aluminum hydroxide precipitate formed after a few seconds of hydrolysis. These experimental data supported the coagulation mechanisms proposed by previous researchers (Amirtharajah and Mills, 1982; Hahn and Stumm, 1968). In this study, it was shown that the formation of polymeric hydrolysis species took place within 0.1 seconds and that, ideally, coagulants should be homogeneously dispersed within this time frame. However, in the same study the authors compared five full-scale rapid mix systems and found that homogeneous dispersion of coagulant within 0.1 seconds was not realistic, an observation made by other researchers (Dittman et al., 1988). Therefore, the conclusion was that the highly positively-charged polymeric species played a minor role in the charge neutralization of natural particles.

ROLE OF ALUMINUM HYDROXIDE SOLS

A majority of the evidence in the literature, however, suggests that, under water treatment conditions, the positively-charged, insoluble aluminum hydroxide sols are primarily responsible for particle destabilization during coagulation by adsorbing to the surface of the oppositely-charged, much larger natural colloids and entering into an electrical bond (Wang et al., 2002). At higher pH levels (e.g., 7.5) and coagulant doses, this interaction between voluminous aluminum hydroxide precipitates and natural colloids is known as sweep coagulation (Amirtharajah and Mills, 1982; Amirtharajah and Tambo, 1991).

Licskó (1997) showed that, for alum doses $>10^{-2}$ M and in the pH range of 4.0 to 5.0, where the stability of positively-charged polynuclear species is high, the alum dose required to change the zeta potential of natural particles from a river source was significantly greater than in the pH range 6.0 to 6.5, where aluminum hydroxide sols are formed within 1 to 2 seconds. In the pH range 6.6 to 7.8, it was shown that greater than 98 percent of the applied alum was present as insoluble hydroxide within approximately 10 seconds. The author emphasized that the positively-charged sols themselves are relatively short-lived species (1 to 10 seconds) that readily aggregate into $Al(OH)_{3\ (S)}$ microfloc. This sol-sol interaction was viewed as an ineffective use of coagulant. Additional experiments confirmed that "aged" aluminum hydroxide flocs formed 60 seconds after alum addition did not reduce turbidity and were significantly less effective for charge neutralization than the species formed within a few seconds of alum addition.

Dentel and Gossett (1988) also concluded that the formation and adsorption of positively charged precipitates (sols) to suspended particles achieves charge neutralization. They supported this conclusion by comparing experimental electrophoretic mobility data to aluminum solubility diagrams. This comparison showed that charge neutralization did not take place until aluminum hydroxide solubility limits were exceeded, a mechanism that had been suggested by prior researchers (Mackrle, 1962; Hall, 1966). In another study by DiTommaso and Van Benschoten (1996), alum was dosed into a river water (dissolved organic carbon (DOC) = 5.1 mg/L) and allowed to react for 5 seconds under rapid mix conditions at low pH before the addition of base to achieve the target final pH, which ranged from 4.5 to 7.5. For three different alum doses, the most effective particle removal was achieved at pH = 6 to 7, the range in which the aluminum would be converted completely to insoluble aluminum hydroxide sols within a few seconds. The 5-second low-pH contact time between the more positively-charged soluble aluminum species and the suspended particles did not appear to be significant. UV_{254} removal, although less adversely impacted at lower pH levels, closely followed particle removal in these experiments.

EFFECT OF COAGULATION pH ON PARTICLE REMOVAL/FLOC DEVELOPMENT

In terms of the impact of pH and alum dose on coagulation and solid-liquid separation processes, a majority of research has focused on settled water data. These results have shown that particle removal through the sedimentation process improves significantly as pH is increased. Tseng et al. (2000) observed that turbidity removal in jar tests improved directly as pH was increased incrementally from 5.7 to 7.2. In this work, alum doses were increased with pH to maintain a constant zeta potential. Although a definitive explanation for the increased turbidity removal with increasing pH was not provided, the authors noted that the logarithm of percentage turbidity reduction exhibited a rough linear correlation to the mass of $Al(OH)_{3\ (S)}$ formed. Likewise, in the bench-scale alum coagulation experiments conducted by DiTomasso and van Benschoten (1996), the most effective turbidity removal was achieved at the highest alum dose and pH levels (6.9 to 7.4). Budd et al. (1996) also observed in jar test experiments that the optimum pH range for particle removal when using alum was 7.2 to 7.6 for cold water.

Matsui et al. (1998) used a photometric dispersion analyzer (PDA) to perform a dynamic analysis of coagulation mechanisms during bench-scale experiments. They reported that the time required for particle destabilization and the onset of floc formation increased as pH and alum dose decreased. The output from the PDA, called the flocculation index (FI), is a sensitive indicator of the state of aggregation of a suspension (Gregory, 1984) and has been directly related to particle size (Ohto et al., 1993). A rapidly increasing FI following alum addition, which was observed by Matsui et al. (1998) at higher pH levels, is indicative of optimum particle destabilization conditions. In another study, in which the PDA output was used to control microfloc size and thus settled water quality in a pilot plant, coagulation was optimized at pH = 7.4 (Tambo et al., 1997).

IMPACT OF COAGULATION pH ON FILTRATION

Although it seems clear from the literature, as well as from bench- and pilot-scale data to be presented here, that coagulation at relatively high pH and alum dose conditions enhances turbidity removal in the sedimentation process, regardless of zeta potential, it is less clear whether these benefits extend to filtration. Although several excellent studies have focused on modeling the theory of deep-bed filtration using controlled laboratory systems, filtration research focusing specifically on pH and zeta potential effects is not well represented. Edzwald and Tobiason (1999) presented jar test results and pilot-scale filtration data in a study which examined the effect of pH on multiple objective coagulation processes (i.e., balancing coagulation pH and dose with particle, DOC, and total tri-halomethane formation potential

(TTHMFP) removal). This work focused on low-alkalinity waters that require base/alkalinity addition for pH control during coagulation. The authors emphasized the need for strict pH control to achieve optimum coagulation. For colder waters, a coagulation pH near 7 was recommended in order to maximize particle removal without compromising NOM removal. Although it was shown, for example, that less alum is required to remove similar amounts of DOC at pH = 5.5 vs. 6.1, at optimized doses there was little difference in DOC removal over the pH range of 5.5 to 6.6.

In most cases authors have optimized filtration conditions based on surface charge (electrophoretic mobility or zeta potential) and not necessarily pH. O'Melia and Stumm (1967) concluded that "no unique relationship exists between pH and filtration efficiency." In other words, there is not one pH level that will produce optimum filtration results for all coagulation conditions. In laboratory-scale filtration experiments conducted with Fe^{3+} dispersions the authors showed that the optimum pH for filtration varied with several variables, including the age of the dispersion, the presence of complex-forming ligands other than OH^-, and ionic strength. Overall, filtration efficiency was greatest at pH = 7.0 (compared to pH = 5.0 and 9.5), which is near the isoelectric point of Fe_2O_3, but the authors emphasized that these results applied only to the specific type of particle dispersion examined. Darby and Lawler (1990) modeled the removal of different size particles, filter ripening, and head loss accumulation at pH = 2, conditions that were optimized for filtration on the basis of electrophoretic mobility (e.m.) measurements. The e.m. range used was -1.1 to +1.3 (μm/s)/(V/cm). Elimelech and O'Melia (1990) performed laboratory-scale filtration experiments at a constant pH of 6.7, a typical level used in water treatment. Surface electric potential, particle removal, and filter attachment efficiency (α) were found to vary directly with electrolyte concentration. For the small particle sizes examined (<1 μm), removal and α increased as surface potential decreased (became less negative). Although constant pH conditions were used in this study, the changes in surface potential observed over a range of electrolyte concentrations are analogous to surface chemistry changes during coagulation. In both cases, the repulsive energy barrier of a suspended particle is reduced, increasing attachment efficiency to filter media.

It appears that discussions of the impact of pH on filtration should focus on its effect on suspended particle surface chemistry, the most dominant variable affecting the attachment of particles to filter media (O'Melia, 1985). Surface chemistry is usually characterized by the effective net surface charge, or zeta potential, of particles. Changes in coagulation pH, for a constant coagulant dose, are accompanied by changes in zeta potential, i.e., zeta potential becomes more positive as pH decreases and more negative as it increases. To achieve effective removal in filtration it is essential to charge neutralize, or destabilize, the particles in suspension through optimized pre-treatment processes (Chipps et al., 2001; Lawler and Benjamin, 2002). In doing so, the attachment of suspended particles to filter media upon collisions between the two is maximized. The importance of this attachment mechanism and proper particle destabilization, as well as the relative unimportance of the straining mechanism in filtration, has been demonstrated in several studies (e.g., Habibian and O'Melia, 1975; Cleasby et al., 1992). In their classic filtration theory describing the relative importance of different particle-to-media grain transport mechanisms and subsequent removal efficiencies, Yao et al. (1971) concluded that there are many opportunities for particle-media grain collisions and that a high removal percentage should be achieved as long as proper destabilization conditions are maintained. In their equation describing particle removal, the attachment coefficient, α, was used. A value of $\alpha = 1$ implies complete destabilization while $\alpha = 0.1$, for example, indicates poor particle destabilization.

Figure 2.1 illustrates the dramatic impact of attachment efficiency on clean bed particle removal for the Yao et al. (1971) model. The removal of smaller particles (<1 μm diameter), in particular, is extremely negatively impacted by poor destabilization conditions. Although this model appears to predict inadequate particle removal for a water treatment plant, it should be noted that clean bed conditions are assumed, i.e., no ripening has occurred. Elimelech and O'Melia (1991) showed that α was directly related to the concentration of a destabilizing electrolyte and electrokinetic potential, which was calculated from electrophoretic mobility.

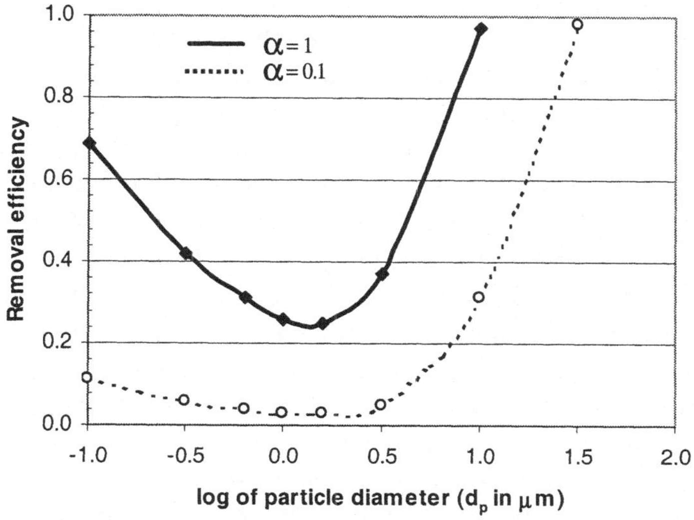

Figure 2.1 Effect of attachment efficiency on overall filter removal efficiency (data from Lawler and Benjamin, 2002)

The overall effect of changes in pH, for a constant alum dose, is to change the surface charge or zeta potential of suspended particles, therefore changing α. To maintain a relatively constant zeta potential and α, alum doses must be adjusted with pH. Coagulation at relatively high pH levels, for example, requires correspondingly high alum doses due to the increase in the charge density of NOM present as well as the decrease in positive charge of the aluminum hydroxide sols formed within seconds of alum addition (Edzwald and Van Benschoten, 1997).

INFLUENCE OF COAGULATION pH ON NOM REMOVAL

A few general conclusions can be drawn from previous research that has examined NOM removal during alum coagulation: 1) there is a sorbable fraction of NOM, generally characterized by higher molecular weight and hydrophobic tendencies, that is amenable to removal through coagulation and a non-sorbable fraction that can not be removed through practical methods (Edwards, 1997; Randtke, 1988; Croue et al., 1993), and 2) the dominant removal mechanism in the pH range of 5.5 to 8.0 is by adsorption to aluminum hydroxide surfaces (Hundt and O'Melia, 1988; Edwards, 1997). Therefore, over the pH range of interest to this study (6.0 to 7.5), it seems reasonable to assume that equivalent NOM removal can be attained if optimized alum doses are used. Although the conventional notion is that NOM removal is significantly improved at lower pH, it appears from the literature that a more accurate statement is that effective NOM removal requires lower alum doses at low pH than at high pH (Edzwald and Tobiason, 1999). Edzwald and Van Benschoten (1997) showed that equivalent removal is possible at pH = 7.0 (compared to pH = 5.5) if doses are increased proportionally to neutralize the increased negative charge density of NOM. The authors noted that the charge density of the particular aquatic fulvic acid used in their experiments increased from 7.5 to 10 μeq/mg C when pH increased from 5.5 to 6.5, while the alum dose required to provide the same amount of positive charge over that range increased by a factor of three.

Other studies have demonstrated that NOM can be removed effectively over a relatively wide pH range. Hundt and O'Melia (1988), for example, showed that, for a constant alum dose, removal of some types of fulvic acids was influenced by pH while for other types high removal rates were observed over a wide pH range. Overall, they concluded that a constant filtered removal percentage (80%) of fulvic acids could be achieved while pH was increased from 6.0 to 8.0, if alum doses were increased accordingly. Edwards (1997), in developing a semiempirical predictive model for DOC removal, noted that inclusion of a pH-dependency factor in the sorption constant did not improve model fits. It was suggested that the affinity of DOC for oxide surfaces was not a function of pH. Exall and Vanloon (2000) did not directly examine the effect of pH on NOM removal, but used a relatively high pH (≈8.0) in their study of organic matter removal by different aluminum-based coagulants. High removal rates of humic acids (almost 75 percent) were observed at optimum alum doses. Bell-Ajy et al. (2000) compared conventional (baseline) and optimized coagulation for NOM removal. The authors examined 16 tests sites, four of which used alum, and concluded that the use of lower coagulation pH (5.5 to 6.5) optimizes NOM removal. However, at the four sites in which attempts were made to optimize alum coagulation, results were mixed. At two of the sites, reducing the coagulation pH from 7.0 to 6.2 and 6.9 to 5.8 had an insignificant effect on TOC removal. At the other two sites, TOC removal increased after the pH was depressed, but the alum was not changed, suggesting that the original baseline process was non-optimum in terms of zeta potential.

TWO-STAGE COAGULATION PROCESSES

Very few studies were available documenting two-stage coagulation processes. Carlson and Gregory (2000) and Carlson et al. (1996), in work that led to the present project, performed preliminary experiments using high-TOC runoff waters. The authors hypothesized that NOM was effectively restabilizing natural particles that had been destabilized by charge neutralization. In jar tests and pilot-scale filtration experiments the two-stage process was often the most effective for turbidity removal. Billica and Gertig (2000) also reported that two-stage processes decreased settled water turbidity significantly and appeared to enhance filtration performance as well.

Wahlroos (1991) reported the use of a two-stage coagulation process in which iron (III) sulfate-chloride was added at a pH of 4.8 to 5.1, followed by pH adjustment and then a second stage of iron coagulation at a pH of 8.0 to 8.5. The two-stage method was initially developed and tested in bench- and pilot-scale experiments before being implemented at a full-scale facility. The process increased filter run times and increased removal of taste-and-odor compounds from 50 to 90 percent. Removal of chemical oxygen demand (COD) was also increased when the coagulation process was changed from one to two stages. The first coagulation process was intended for removal of humic substances and the second process was intended for particle removal and precipitation of metals.

Another study (Garrotte et al., 1995) describes effective treatment of a tannery waste stream, containing high concentrations of biochemical oxygen demand (BOD), COD, and sulfides, by a two-step coagulation process. The authors, using COD reduction to gauge the effectiveness of the process, found that two similar coagulation stages were far more effective than a single stage, but that three stages were only marginally more effective than two. Although the use of two-stage coagulation processes are not well-documented, these studies, in addition to preliminary bench and pilot-scale experiments, have indicated that the process may improve settling and filtration performance, particularly for high-NOM raw waters.

HYPOTHESIZED REMOVAL MECHANISMS

Although it seems clear from the literature that higher coagulation pH conditions, and thus higher alum doses, enhance particle removal in the sedimentation process, the precise reasons for these results are less clear. However, given the apparent dominance of the sol-adsorption mechanisms in the particle destabilization and NOM removal processes, it appears that higher optimized alum doses may be more effective by more uniformly distributing positive charges on particle surfaces. In the pH range of 6.0 to 8.0, dissolved organic molecules adsorb rapidly to newly-formed aluminum hydroxide sols, essentially restabilizing them and preventing their adsorption to particles in solution. Therefore, a certain fraction of the sols are used to meet the coagulant "demand" exerted by NOM. The remaining fraction is available for destabilization of particles as well as sol-NOM complexes. At low pH, this fraction is relatively low and particles are destabilized by fewer but more highly-charged sols. While the net surface charge is zero, distribution of that charge may be less consistent than at higher pH levels. Ultimately, an uneven surface charge distribution may reduce the probability of successful collisions (Amal et al., 1992), hindering the development of floc suitable to solid-liquid separation processes, i.e., large floc. When the pH is raised, the sol demand from the hydrophobic NOM fraction will increase somewhat due to increased negative charge density, but a higher concentration of aluminum hydroxide is available for adsorption and charge neutralization, resulting in a more even distribution of charge. Elimelech and O'Melia (1990) also discussed distribution of surface charge as a possible explanation for discrepancies between theoretical and experimentally observed attachment coefficients, but no experiments were conducted to directly examine it.

CHAPTER 3
MATERIALS AND METHODS

GENERAL APPROACH

The general experimental approach was to first illustrate the effect of different coagulation scenarios on particle/NOM removal and microfloc formation kinetics in a controlled laboratory environment. In these experiments, alum doses were optimized based on zeta potential over the pH range of 5.5 to 7.5. This bench-scale phase was performed in the environmental engineering laboratories at Colorado State University (Fort Collins, Colorado). In the second phase, pilot-scale experiments were conducted to correlate bench-scale findings with granular media filtration results. The pilot-scale experiments were executed at the pilot laboratory of the City of Fort Collins (Colorado) Water Treatment Facility. Filtration runs were conducted at pH = 6.0, 6.7, and 7.5 using two different surface water sources: 1) a mountain river during spring runoff (high-NOM/low alkalinity, highly variable water quality), and 2) a reservoir characterized by low-moderate NOM levels and generally stable water quality (alkalinity = 35 to 40 mg/L as $CaCO_3$). During the reservoir water experiments, a two-stage alum addition process was applied in addition to the three single-stage processes. Other than acid/base addition for pH control, no coagulation chemicals other than alum were used in the bench- and pilot-scale experiments, nor were pre-oxidants applied.

BENCH-SCALE EXPERIMENTS: PDA

Microfloc formation kinetics were monitored with a photometric dispersion analyzer (PDA). The output from the PDA, or flocculation index (FI), corresponds to particle size and has been shown to be a sensitive indicator of the state of aggregation of a particle suspension. The method was originally developed by Gregory (1984) and has been described in detail by several researchers who used it to monitor microfloc development during coagulation (Matsui and Tambo, 1991; Ohto et al., 1993; Tambo et al., 1997; Huang and Chen, 1996). It is based on the fact that turbidity measurements, i.e., measurements of average transmitted light, of a flowing suspension fluctuate about the mean value. The intensity of transmitted light, which is converted to a voltage signal, will change continually because of fluctuations in the number and size of particles in the path of the beam. The FI is defined as the ratio of the root mean square of the fluctuations to the mean level. The voltage signal is split into its average (DC) and fluctuating (AC) components, which are used for calculating the FI. The most practical use of the FI is that it shows a significant increase as particles aggregate, thus allowing a dynamic assessment of coagulation mechanisms (Matsui et al., 1998).

Much of the early focus of this project was placed on developing a protocol for using the PDA (provided by Hach Co.) as well as understanding the FI data obtained from it. The PDA experimental protocol was as follows: a 2.3 L sample of water was added to a standard jar test square container and the pH was adjusted according to the experimental conditions. A laboratory-scale mixer (IKA®-Works Eurostar model, rpm range = 50 to 2,000) was lowered into the container and the sample was mixed at 300 rpm ($G \approx 320\ sec^{-1}$). At t = 0 the FI data logging using Labview software was started. Baseline FI data were collected for 150 seconds prior to coagulant addition. Coagulant was added at t = 150 seconds—injected with a micropipette—and the sample was mixed at 300 rpm for an additional 15 seconds, after which time the mixing intensity was decreased to 60 rpm. The sample was then mixed continuously at 60 rpm for an additional 12 minutes while the FI data were collected. At the end of the experiment the data logger was turned off and the file was automatically saved to the computer hard drive. A schematic diagram of the PDA system used in this study is provided in Figure 3.1. The peristaltic pump (flow rate = 46 mL/min.) was placed downstream of the flow-through cell to avoid disruption of the floc. The time lag between the 2-L beaker and the PDA was 3 to 4 seconds.

In the early stages of the PDA work, many experiments were performed using synthetic waters containing relatively high DOC concentrations. In this bench-scale phase, FI data were correlated with settled water results and the PDA was shown to be an effective indicator of overall process performance. As the project progressed, the importance of optimizing alum doses in terms of zeta potential for the different coagulation scenarios became clear. During the final phase of PDA experiments, therefore, emphasis was placed on using optimized alum doses in order to evaluate the different coagulation scenarios. Two waters were used during this stage, one synthetic and the other a natural surface water from a reservoir. For the synthetic water, FI data were correlated with settled water particle and DOC removal from jar tests. For the reservoir water, FI data were correlated primarily with pilot-scale filter effluent quality (DOC and particle counts). The water quality characteristics of concern for both sources are given in Table 3.1. The DOC concentrations of 3.3 and 8.8 mg/L for the synthetic water were chosen in order to approximate the levels in the reservoir and spring runoff waters, respectively, used in the pilot-scale phase. For the synthetic water, the TOC and DOC concentrations measured using the UV/persulfate oxidation method (Sievers 800 TOC Analyzer), were equal. DOC samples were filtered through a 0.45 μm membrane. Suwannee River NOM (International Humic Substance Society—48% carbon by weight) was dissolved in de-ionized water to achieve the target DOC

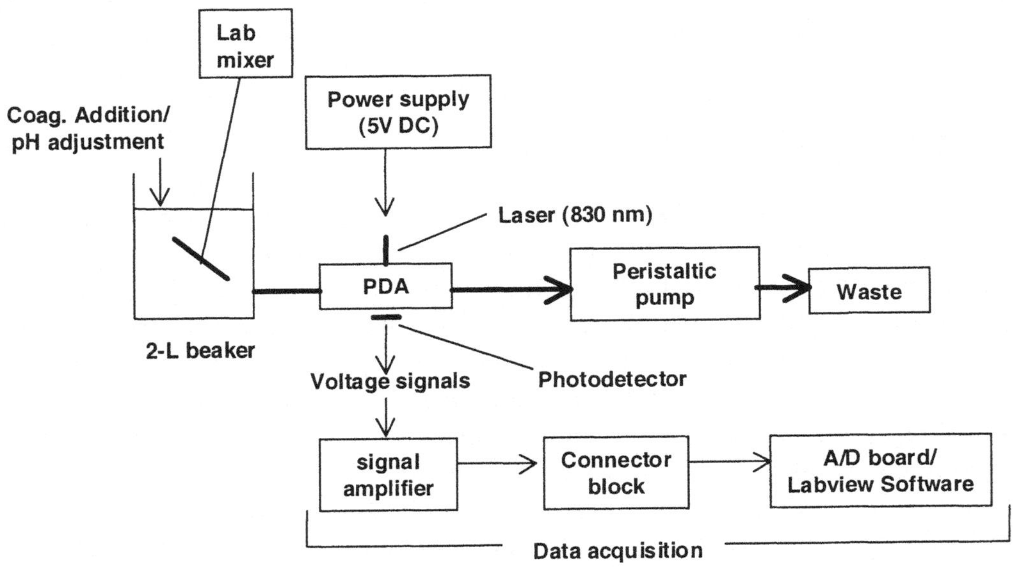

Figure 3.1 Schematic diagram of bench-scale PDA apparatus and data-acquisition system

concentrations. Sodium bicarbonate and calcium chloride stock solutions (10^{-3} M) were used for alkalinity and hardness addition, respectively. Turbidity was adjusted by the addition of small volumes of a kaolin stock suspension. The dose to achieve 2 ntu was approximately 10 mg/L, from 1 g/L stock.

Table 3.1
Water quality characteristics of synthetic and natural (reservoir) sources used in bench-scale experiments

Constituent	Reservoir water	Synthetic water (DI)
TOC (mg/L)	3.5	3.3, 8.8
Alkalinity (mg/L as $CaCO_3$)	35	20
Turbidity (ntu)	10 - 20	2
Hardness (mg/L as $CaCO_3$)	30	25
pH	7.5 (ambient)*	5.5, 6.0, 6.7, 7.5*

* Adjusted according to requirements for each experiment.

For the synthetic water experiments, it was necessary to determine the optimized alum doses at each pH/DOC level. Prior to the addition of alum, an appropriate dose of acid (2.5 N HCl) or base (2.5 N NaOH), determined by titrations, was added to achieve the target coagulation pH of 5.5, 6.0, 6.7, or 7.5. A programmable 2-L jar test apparatus was used (Phipps & Bird) to provide a four-stage coagulation/flocculation mixing process: 1) 20 sec. rapid mix at $G \approx 300$ sec.$^{-1}$ (300 rpm), 2) 15 min. slow mix at $G \approx 48$ sec.$^{-1}$ (48 rpm), 3) 15 min slow mix at $G \approx 32$ sec.$^{-1}$ (32 rpm), and 4) 15 min. slow mix at $G \approx 18$ sec.$^{-1}$ (18 rpm). Following the mixing cycle, a 15-min. settling period was provided, after which samples were collected and analyzed for residual turbidity (Hach 2100), TOC/DOC, and zeta potential (Zeta-Meter 3.0^{+}). The sampling port was located 11 cm below the water surface. For each pH level, a range of 6 to 10 alum doses was applied and all experiments were performed in duplicate. Dose optimization was based on zeta potential. Optimum doses were defined as those that achieved a zeta potential closest to zero *and* within the range of 0.0 to ±2.0 mV. The purpose of these experiments was to characterize alum dose, turbidity removal, and DOC removal as a function of pH when zeta potential is optimized. The use of DOC, rather than TOC, removal for process optimization has been suggested previously (Edzwald and Tobiason, 1999) and is based on the assumption that conventional filtration water treatment facilities will remove particulate organic carbon.

PILOT-SCALE EXPERIMENTS

The Fort Collins pilot facility consists of three 8 gpm conventional treatment process trains which can be operated in parallel. The simultaneous use of multiple process trains allowed a direct comparison of the impact of different coagulation conditions on filtration performance. For the rapid mix process, both in-line static mixers and standard backmix-type units were available. For the single-stage coagulation processes, alum was added into the backmix basins (detention time = 15 sec., mixer speed = 1,000 rpm, G value $\approx$ 1,000 sec.$^{-1}$). For the two-stage alum addition processes, the first alum dose was injected into a 2 in. PVC line containing an in-line static mixer, about 25 seconds upstream of the backmix basin and the second alum dose.

Each pilot train included a three-stage, tapered mixing energy flocculation basin (total detention time = 45 min., G values ≈ 50, 30, and 18 $sec.^{-1}$), Lamella plates settling basin (hydraulic loading rate = 0.34 gpm/ft^2), and a dual media filter (30 in. anthracite, effective size = 0.95 mm; 10 in. sand, effective size = 0.50 mm). The filtration rate used for all experiments was 4 gpm/ft^2. Filtration performance was evaluated based on filter effluent particle counts (total > 2μm/mL; Met One particle counter) and TOC concentration. TOC samples were collected after the filters reached steady-state conditions. Each pilot train was equipped with on-line pH meters (Great Lakes Instruments) for raw and post-rapid mix water, as well as on-line turbidimeters for raw, settled, and filtered water (Hach 1700D). Alum doses at each pH level were optimized based on zeta potential (automated Malverne Instruments Zetasizer 2000). The original target zeta potential for coagulation was 0.0 mV, but preliminary experiments clearly demonstrated that filtration conditions were optimized at slightly positive zeta potentials. The coagulation target, therefore, was adjusted to +2.5 ± 1.5 mV. Reasons for this z.p. target are discussed below.

Spring Runoff Experiments

Filtration runs were performed during the spring runoff season in May through June 2001, using water from the Cache la Poudre River, which originates at the Continental Divide in the Rocky Mountains outside of Fort Collins, Colorado. Only single-stage alum addition processes were used during the runoff experiments. Figure 3.2 provides turbidity, TOC, and alkalinity data for the river during the testing period. In terms of treatment, the spring runoff water quality is typically challenging and requires blending with a more stable reservoir source to meet finished water quality goals. The TOC increases from a non-runoff baseline of about 1 mg/L to 7 to 12 mg/L, while the alkalinity falls to approximately 10 mg/L as $CaCO_3$. The high NOM content requires the use of relatively high alum doses, which in turn necessitates base addition for pH control. Two target coagulation pH levels, 6.1 ± 0.1 and 7.4 ± 0.1, were used for the runoff experiments, with alum doses optimized at approximately 37 and 97 mg/L, respectively, for a majority of the period. To control pH, a 200 g/L sodium hydroxide solution was injected at an appropriate flowrate upstream of the alum addition point.

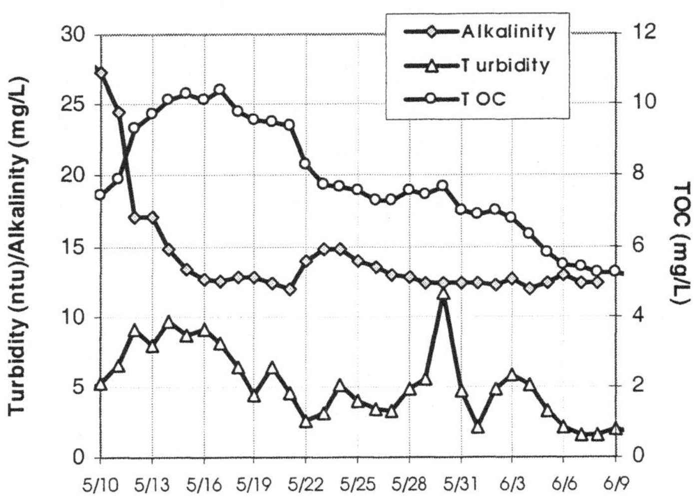

Figure 3.2 Poudre River water quality during spring runoff, 2001 (pH = 6.8 to 7.0, T = 6 to 8°C)

Reservoir Water Experiments

For the reservoir source, the quality of which is relatively constant throughout the year (Figure 3.3), filtration experiments were completed in October to November 2001. The single stage alum target coagulation pH levels were 6.0, 6.7, and 7.4. The pH = 6.0 conditions were achieved by diffusing carbon dioxide gas ($CO_{2\,(g)}$) into a PVC contactor, thirty minutes upstream of the alum addition point. Optimized alum doses (i.e., zeta potential ~ 2.5 mV) at this pH level ranged from 15 to 18 mg/L. The pH = 6.7 process required no acid or base addition (ambient conditions) prior to alum addition. Optimized alum doses were 21 to 23 mg/L. For the pH = 7.4 process, NaOH solution was added approximately 3 minutes upstream of alum. For this process, alum doses were optimized at 39 to 41 mg/L.

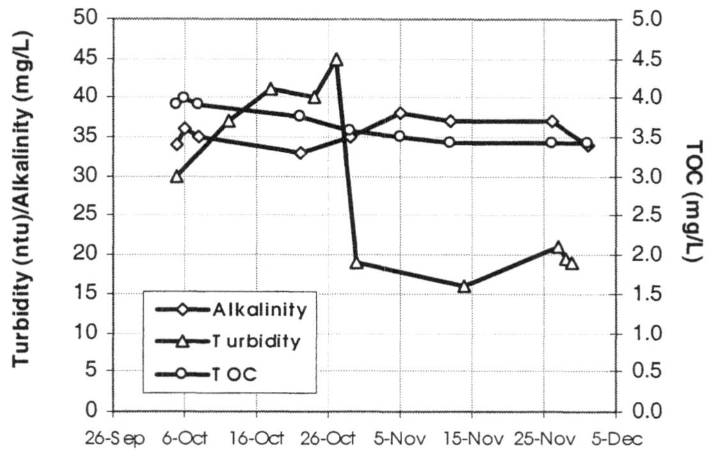

Figure 3.3 Horsetooth Reservoir water quality, Fall 2001 (pH = 7.4 to 7.5, T = 7 to 9°C)

Two-stage alum addition

A two-stage alum addition process was also used during this phase of the pilot-scale experiments. The pH for the first stage was approximately 6.0 and the pH for the second stage was approximately 7.4. The alum dose for each stage was optimized based on z.p. Carbon dioxide gas was added prior to the first alum dose (A1) to achieve pH = 6.0 and sodium hydroxide was added simultaneously with alum in the second stage (A2). A1 and A2 were approximately 16 and 44 mg/L, respectively. During the two-stage runs, A1 was injected at an in-line static mixer approximately 25 seconds upstream of the rapid mix basin into which A2 and sodium hydroxide were injected.

CHAPTER 4
RESULTS AND DISCUSSION

BENCH-SCALE EXPERIMENTS

In the first set of bench-scale experiments using the synthetic waters, alum doses were optimized based on zeta potential for two TOC concentrations over a coagulation pH range of 5.5 to 7.5 (Figure 4.1). Optimized doses were defined as those that resulted in a z.p. closest to zero and were within the target range of 0.0 ± 2.0 mV. As expected, optimum doses increased with pH due to the increasing negative charge density of NOM and decreasing positive charge of aluminum hydroxide sols.

Settled water residual turbidity and DOC data corresponding to the optimized alum doses are shown in Figures 4.2 and 4.3, respectively. Residual turbidity increased directly with coagulation pH, with a more dramatic impact observed for the high-TOC water. In both cases, particle removal was optimized at pH = 7.5. DOC removal was maximized at pH = 6.0 but, particularly for the high-TOC water, the impact of pH was relatively minor. These results indicate that the hydrophobic fraction of NOM is readily removable through adsorption to aluminum hydroxide precipitates over a relatively wide pH. The interference of NOM with the formation of settleable floc is also evident, particularly at lower pH levels.

In the case of the high-TOC water, one question that arises is why turbidity removal improves dramatically with increasing pH when the zeta potentials are relatively constant. This situation appears to be due to the physical characteristics of the floc formed at high pH. As discussed earlier, coagulation at higher pH may produce a more uniform charge distribution on the surfaces of suspended particles due to the increased concentration of adsorbable, positively-charged aluminum hydroxide available for this purpose. Under these conditions, the probability of successful collisions may be increased and the development of floc suitable to liquid-solid separation is promoted.

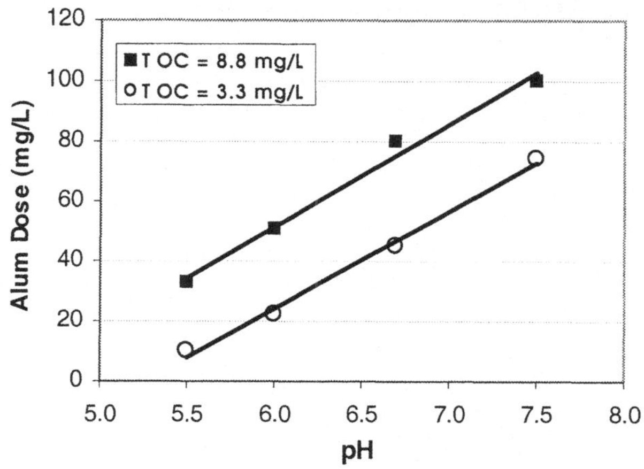

Figure 4.1 Optimized alum dose vs. pH for synthetic water at two TOC concentrations

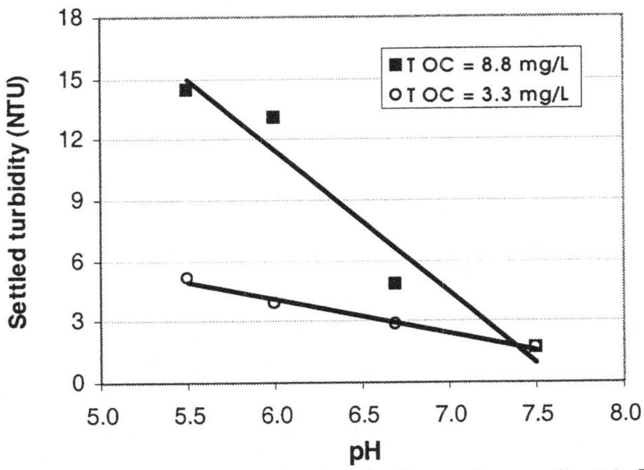

Figure 4.2 Settled turbidity vs. pH for optimized alum doses (initial turbidity = 2 ntu)

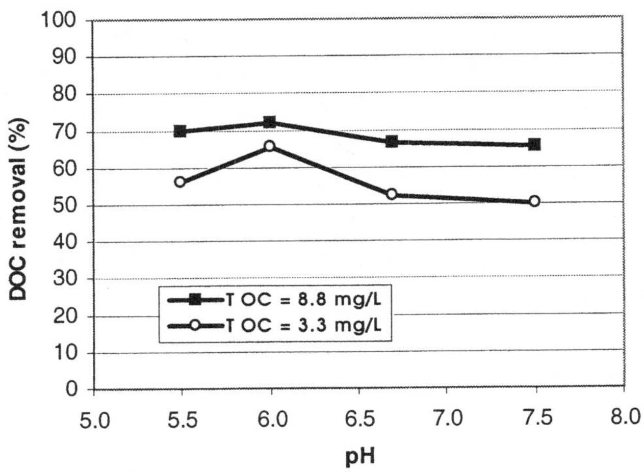

Figure 4.3 DOC removal vs. pH for optimized alum doses

The rate of floc development vs. pH and alum dose is illustrated by results from the PDA in Figure 4.4. The high-TOC synthetic water was used in these experiments, at the same pH levels and optimized alum doses as shown in Figures 4.1 through 4.3. The rate of floc formation clearly increases with pH and optimized alum dose. It should be noted that FI is not simply a function of alum dose and precipitation of $Al(OH)_{3\ (S)}$. As shown in Figure 4.5 (using a similar high-TOC synthetic water), non-optimum alum doses are reflected by relatively small or very slow changes in the FI, even for pH levels at which $Al(OH)_{3\ (S)}$ would form rapidly. In this case the alum dose of 50 mg/L at pH = 7.2 was sub-optimum (z.p. = -7.2 mV).

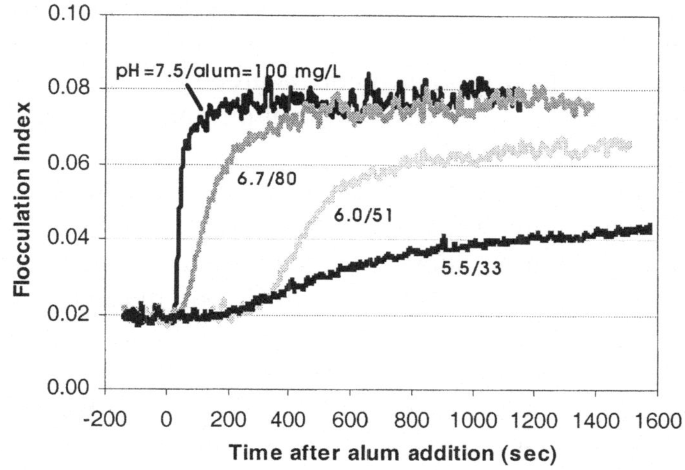

Figure 4.4 FI vs. time for optimized alum doses at four pH levels (TOC = 8.8 mg/L, T = 9°C)

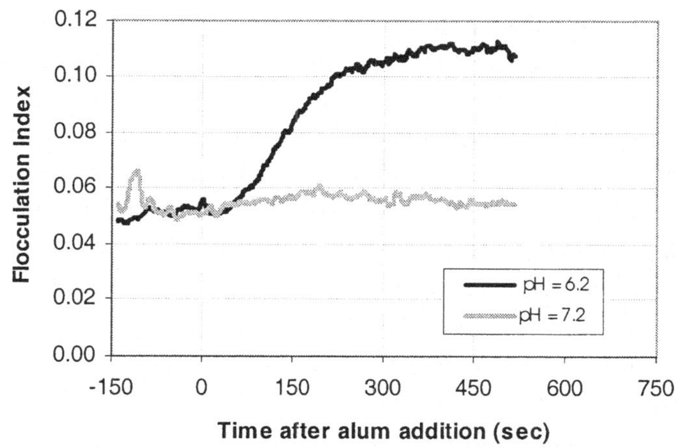

Figure 4.5 FI vs. time after alum addition (50 mg/L) (T = 10°C, TOC = 9.2 mg/L, turbidity = 60 ntu (kaolin added))

FILTRATION RESULTS—SPRING RUNOFF WATER

The initial bench-scale phase of this project demonstrated that higher coagulation pH and alum doses enhance particle removal in the sedimentation process without significantly compromising TOC removal. In the second pilot-scale phase, two sets of filtration experiments were performed using similar coagulation pH/z.p. conditions. In the first set, a total of 17 filtration experiments were conducted from May 25 through June 6, 2001, using the spring runoff water. The general purpose of these experiments was to determine the influence of coagulation pH and z.p. on filtration during runoff and, more specifically, the degree to which their impact on the settling process affected filter effluent quality. In most cases, filter runs were terminated upon particle breakthrough. Although most runs using runoff water were short in terms of typical full-scale treatment, it should be noted that the goal was to isolate the relative effects of coagulation pH and z.p. on alum processes, without interference from other coagulation chemicals. Traditionally, this runoff source has been treated by using significant doses of a cationic polymer in conjunction with alum and blending it with the more stable, moderate-NOM level reservoir water.

Initially, alum dose optimization was based on a z.p. target of zero. However, after several preliminary filtration runs were conducted at different z.p. values, it was clear that a slightly positive z.p. was optimum for the experimental conditions being used. The z.p target was therefore adjusted to approximately +2.5 ± 1.5 mV, but runs were also performed at non-optimum z.p. Due to the limited time period during which this water was available, two pH levels were examined during the runoff experiments, 6.1 ± 0.1 and 7.4 ± 0.1. Overall, filtration results for runoff water showed that excellent filter effluent quality (total particles <10/mL) was attainable at both pH levels, but that higher coagulation pH provides a far broader optimum operating range, in terms of z.p. At low pH, and thus low alum doses, steady-state particle counts increased rapidly as z.p. moved away from the optimum conditions. At high pH, and correspondingly high alum doses, optimum filtration performance was sustained over a wide range of z.p.

Figures 4.6 and 4.7 show the effect of alum dose optimization on filtration performance at pH 7.4 and 6.0, respectively. These experiments were conducted in series using the same pilot train and filter over a four-day span (May 26 through 29, 2001) during which the raw water quality was relatively constant. At pH = 7.4, a significant alum dose increase (and corresponding change in z.p.) affected settled water quality only slightly (settled turbidity decreased by 0.1 ntu) while steady-state particle counts decreased by greater than 50 percent. Even at a highly sub-optimum z.p. of -.5 mV, however, total particle counts were about 11/mL at this pH level. At pH = 6.0, alum dose and z.p. changes had a greater impact on filter effluent quality. At a z.p. of 4.4 and 1.2 mV, steady-state particles were approximately 5/mL and 16/mL, respectively, while settled turbidity was 1.0 and 1.3 ntu, respectively. Although the observed effect of dose optimization on particle removal in the settling process was greater at pH = 6.0 than at 7.4, it was relatively minor compared to the influence on filtration performance. TOC removal at the optimized doses, as indicated on the plots, was only slightly higher at low pH.

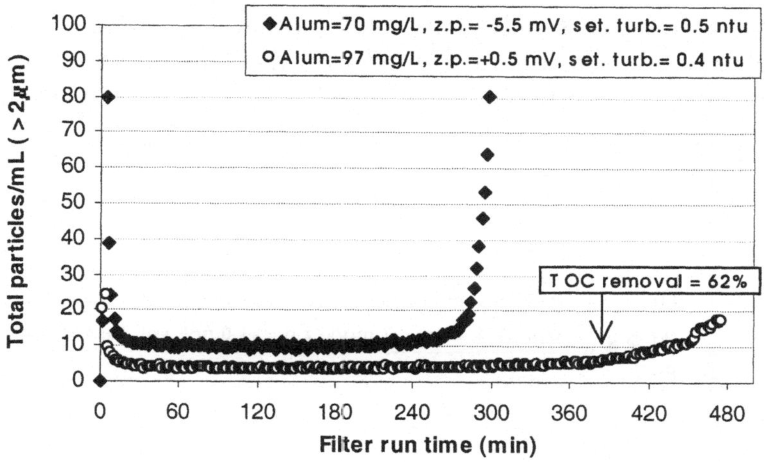

Figure 4.6 Impact of alum dose optimization on filtration performance at pH = 7.4 using runoff water (TOC = 7.5 mg/L)

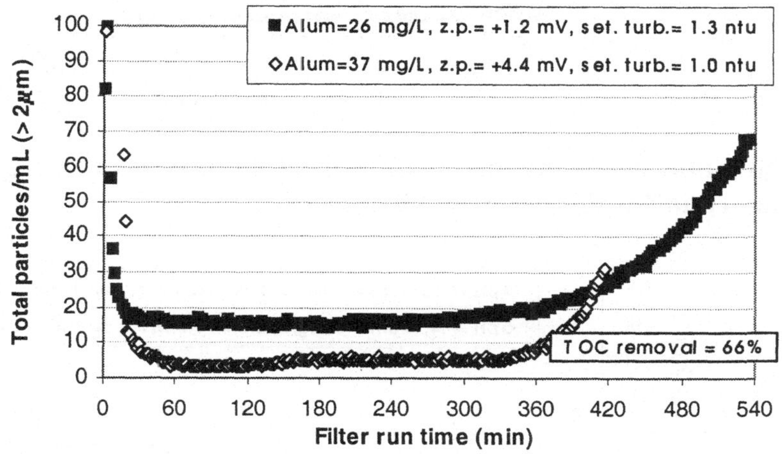

Figure 4.7 Impact of alum dose optimization on filtration performance at pH = 6.0 using runoff water (TOC = 7.5 mg/L)

Figures 4.8 and 4.9 present results from filtration runs performed simultaneously on two parallel process trains. The goal in these runs was to operate at the same z.p. and as close as possible to the target of +2.5 mV. This goal was generally met for the low-pH process, but in practice it was difficult to maintain a positive z.p. at high pH due to small fluctuations in the raw water alkalinity and pH. These data show that the use of higher pH conditions can optimize filtration performance under challenging treatment conditions such as spring runoff. Again, as indicated in Figure 4.9, TOC removal for the two processes was approximately equal.

In Figures 4.10, steady-state settled turbidity and filter effluent particle counts from all the runoff experiments are plotted against the corresponding coagulation z.p. from each run. For the low pH process, alum doses ranged from 26 to 37 mg/L. For the high-pH process, doses ranged from 70 to 97 mg/L. As observed in the bench-scale experiments, turbidity removal in the sedimentation process was consistently greater at higher coagulation pH. Although optimization of filtration performance (i.e., steady-state total particles < 10/ml) was possible for both processes, it was clearly less sensitive to z.p. at high pH. This effect was not due solely to the solids mass loading rate on the filters, as evidenced by the fact that particle counts at low pH were equivalent to those at high pH, but only when coagulation z.p. was in the optimum range of approximately 2 to 4 mV.

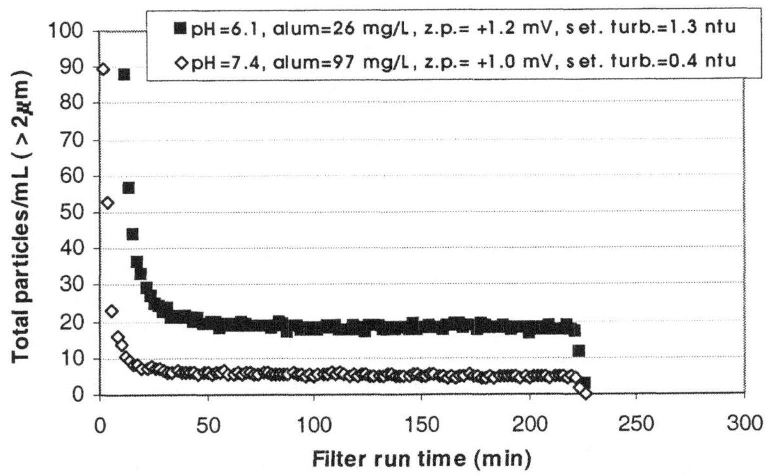

Figure 4.8 Filtration performance for optimized alum doses at coagulation pH = 6.1 and 7.4 using runoff water (TOC = 7.5 mg/L)

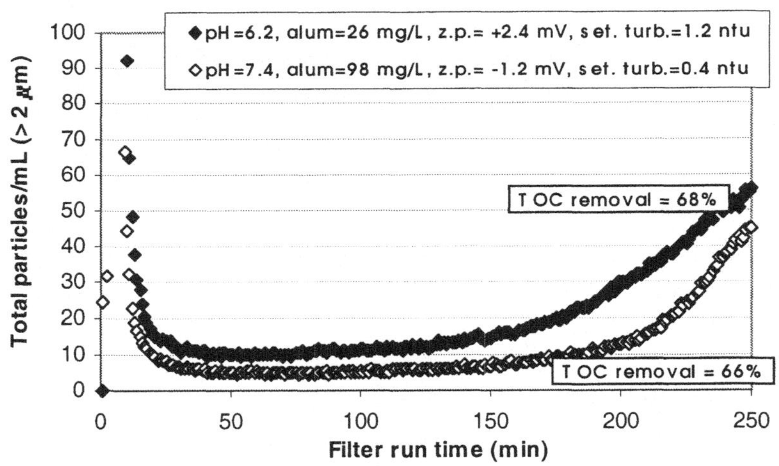

Figure 4.9 Filtration performance for optimized alum doses at coagulation pH = 6.2 and 7.4 using runoff water (TOC = 5.8 mg/L)

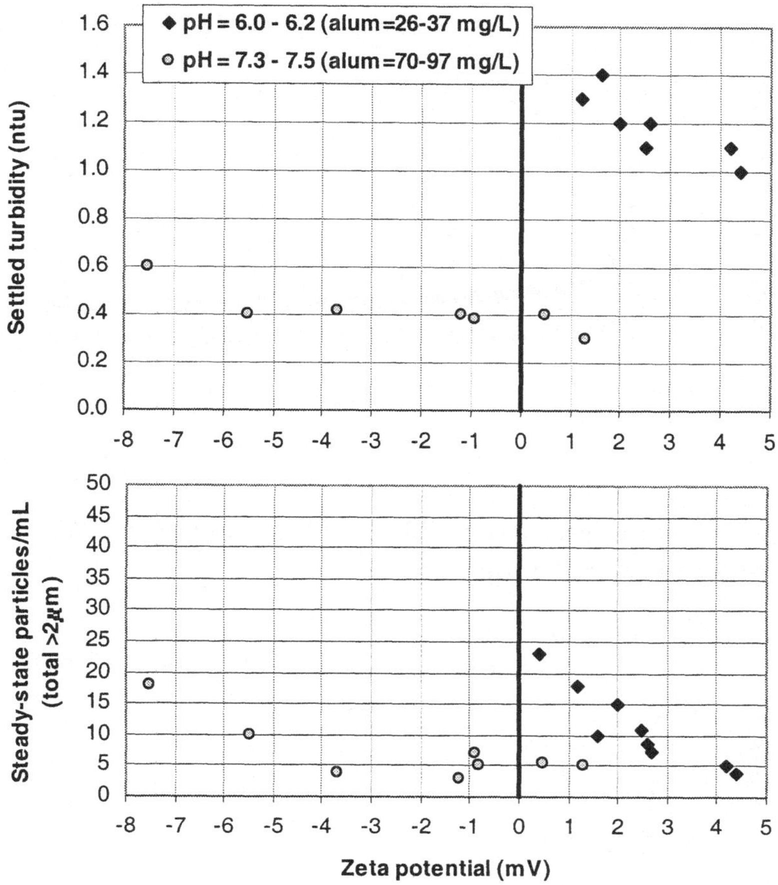

Figure 4.10 Steady-state filter effluent particle counts and settled turbidity vs. z.p. using a spring runoff water, May 25 through June 6, 2001 (TOC 6 to 8 mg/L)

These results appear to indicate that the use of a higher coagulation pH allows process optimization over a broader range of z.p. and alum doses. Although the low-pH filtration performance was not necessarily poor and may be acceptable, depending on site-specific water quality goals, the use of a higher coagulation pH may be particularly advantageous during periods of rapidly changing water quality conditions such as runoff events.

FILTRATION RESULTS—RESERVOIR WATER

In the second set of pilot-scale experiments, a total of 27 filter runs were performed from October 29 to December 3, 2001, using the lower-TOC reservoir water (Figure 3.3). The water quality was relatively constant during this period. Runs were conducted at three pH levels: 6.0, 6.7, and 7.5. Figures 4.11 through 4.13 illustrate the impact of alum dose optimization on filter performance at pH = 6.0, 6.7, and 7.5, respectively. These runs were executed over a 7-day span during which the raw water quality was relatively constant. As observed for the runoff water, alum dose and z.p. changes had a greater effect on filtration performance at lower pH levels.

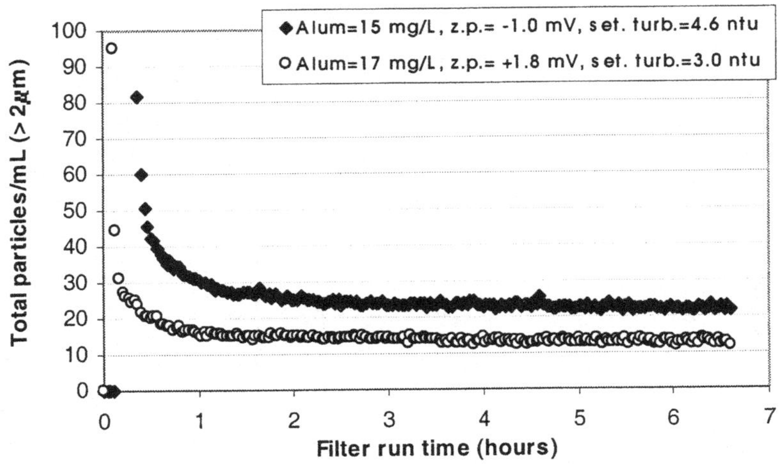

Figure 4.11 Impact of alum dose optimization on filtration performance at pH = 6.0 using reservoir water (TOC = 3.5 mg/L)

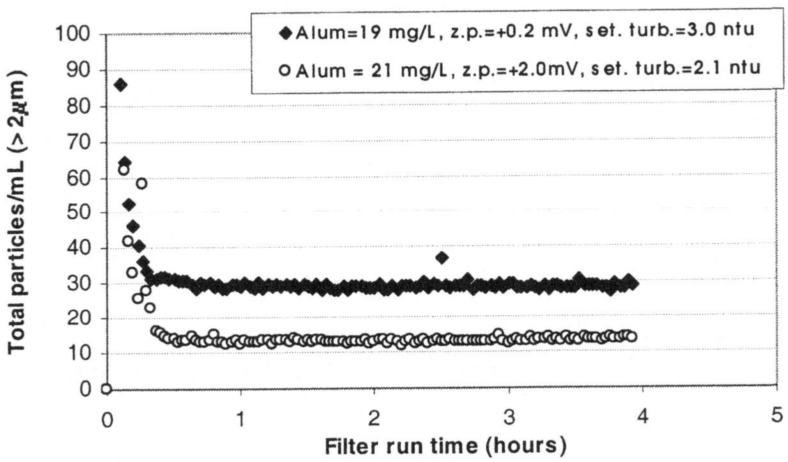

Figure 4.12 Impact of alum dose optimization on filtration performance at pH = 6.7 using reservoir water (TOC = 3.5 mg/L)

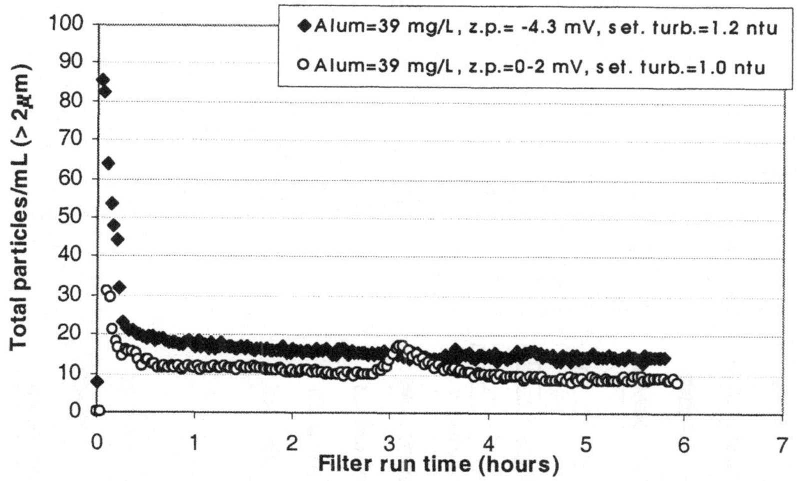

Figure 4.13 Impact of alum dose optimization on filtration performance at pH = 7.4 using reservoir water (TOC = 3.5 mg/L)

At pH = 7.4, for example, total steady-state particles decreased by about 33 percent (from approximately 15 to 10/mL) for a significant change in z.p. (-4 → 1 mV). For a smaller change in z.p. at pH = 6.0 (-1 → 1.8 mV), total steady-state particle counts decreased by about 50 percent. At pH = 6.7, particles decreased by approximately 55 percent for a relatively small change in z.p. (0.2 → 2.0 mV).

As was observed in the runoff water experiments, alum dose optimization had a greater effect on particle removal in the settling process at lower pH levels. At pH = 6.0, for example, settled turbidity decreased from 4.6 to 3.0 ntu for a change in z.p. of +2.8 mV. At pH = 7.4, a change in z.p. of approximately +5.3 mV caused settled turbidity to decrease from 1.2 to 1.0 ntu. Given the fact that TOC removal through the filtration process was approximately equivalent between the three coagulation pH levels (Figure 4.14), it appears that the primary effect of a higher aluminum hydroxide concentration is on particles. As discussed earlier in an effort to understand the bench-scale results, the alum "demand" from NOM, or the concentration of the newly-formed aluminum hydroxide sols to which NOM molecules quickly adsorb and neutralize, appears to be about the same at pH = 6.0 and 7.5. As pH and alum dose increase, a greater concentration of sols is available, and *required* because of the decrease in positive charge density, for adsorption to suspended particle surfaces. A higher concentration of adsorbed aluminum hydroxide on particle surfaces may promote a more uniform surface charge, thus decreasing overall repulsive forces and increasing the probability of particles attaching upon collision. Such a mechanism would likely enhance interparticle bridging and the development of floc suitable for solid-liquid separation processes.

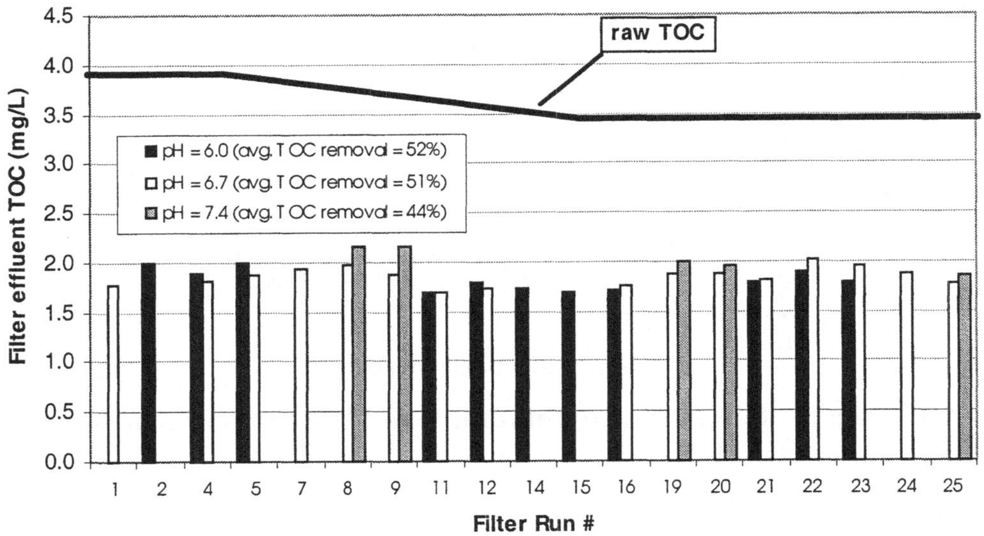

Figure 4.14 Filter effluent TOC vs. coagulation pH for optimized alum doses using reservoir water, November 1 through November 20, 2001 (Raw TOC = 3.5 mg/L)

In Figure 4.15, steady-state settled water turbidity and filter effluent particle counts are plotted against the corresponding coagulation z.p. for all the reservoir filtration runs. The optimum z.p. range for filtration was similar at pH = 6.0 and 6.7 (+1 to +3 mV). At pH = 7.4, the optimum range appeared to be significantly wider than at the two lower pH levels and similar to the observed optimum range using runoff water (Figure 4.10). Likewise, the optimum z.p. range at pH = 6.0 was very similar for the reservoir and runoff sources. A general conclusion that can be drawn from these results is that the optimum z.p. range for filtration is independent of water quality and broadens significantly as pH increases. Again, as was observed in the runoff experiments, this situation does not appear to be due solely to solids mass loading rate. As shown in Figure 4.15, steady-state particle counts equivalent to those attained at pH = 7.4 (settled turbidity ≈ 1 ntu) were often attained at pH = 6.0 and 6.7 (settled turbidity = 2 to 5 ntu) when the z.p. was optimized. To confirm this point, a simple filtration experiment was performed in which an optimum alum dose was used for several hours of steady-state conditions, after which point the CO_2 dose was decreased, therefore increasing the coagulation pH from 6.0 to 6.3 and changing the z.p. from +2.0 to -1.5 mV (Figure 4.16). As shown, settled turbidity decreased from approximately 4.4 to 3.0 ntu while steady-state particle counts increased from approximately 5 to 10/mL.

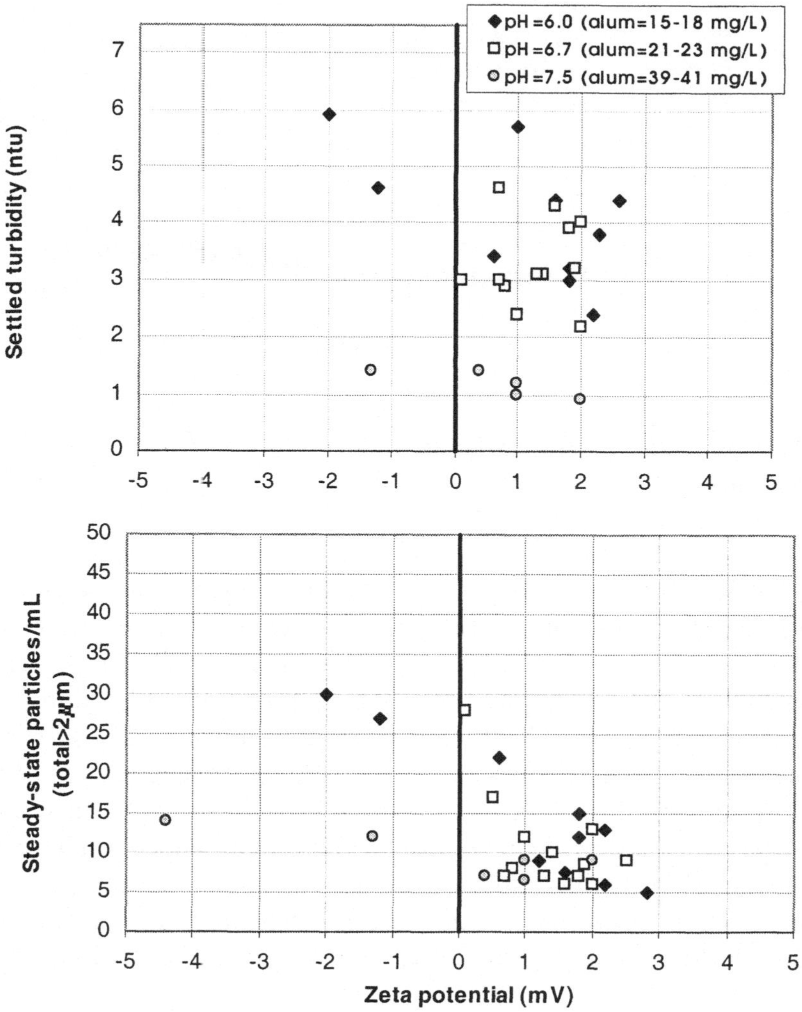

Figure 4.15 Steady-state filter effluent particle counts and settled turbidity vs. z.p. for three coagulation pH levels using reservoir water (Raw TOC = 3.5 mg/L)

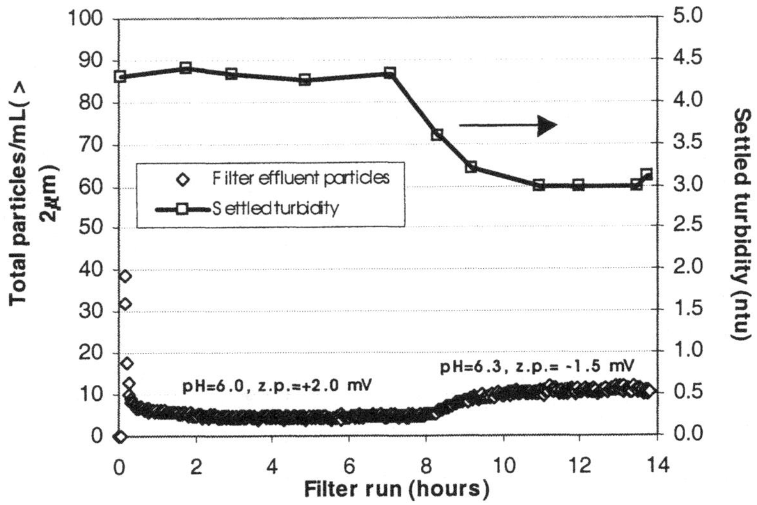

Figure 4.16 Effect of pH/z.p. on settling and filtration for lower pH conditions using reservoir water

Overall, the filtration data indicate that the advantage of higher pH and alum doses may be to increase the filter attachment coefficient, α, of particles over a wider range of z.p. values and coagulation conditions. As hypothesized earlier, given that TOC removal decreased only slightly at higher pH levels (i.e., approximately the same amount of alum hydroxide "demand" was exerted), it appears that a possible mechanism for this result may be a more uniform charge distribution on suspended particle surfaces due to the higher concentration of adsorbed aluminum hydroxide sols. A more uniform surface charge distribution may decrease repulsive energy barriers between particles and promote 1) floc development, and 2) attachment of smaller, non-settleable particles to filter media. Although direct measurements of surface charge distribution were beyond the scope of this study, the data clearly indicate that increasing the coagulation pH increases the rate of floc development, turbidity removal in the settling process, and filtration efficiency.

CORRELATION OF FILTRATION RESULTS WITH PDA DATA

As discussed earlier in Chapter 1 (Research Objectives/Approach), one of the objectives of this project was to evaluate the PDA as an on-line coagulation monitor and indicator of overall process performance, including filtration. In general, it appears that the FI data does correlate well with optimized filtration results and that use of a PDA may be an alternative to jar testing for determining effective coagulation processes. As shown in Figures 4.2 and 4.3, settled water turbidity appears to be directly related to the rate of floc formation, as indicated by the FI output from the PDA. If rate of floc formation is indicative of the average suspended particle surface charge characteristics (i.e., the degree of destabilization achieved), then it seems reasonable to assume that FI would be related to filtration performance as well because floc particles become the primary particle receptors in filters after the initial ripening period (O'Melia, 1985). Figures 4.17 and 4.18 are representative FI curves for optimum and non-optimum alum doses at pH = 6.1 and pH = 6.6 ± 0.1, respectively, using the reservoir water that was used in the filtration experiments. At both pH levels, the rate of FI increase was greatest for the alum dose that was optimum for filtration, in terms of z.p. (see Figure 4.15). At pH = 6.6, doses greater than the optimum of 25 mg/L did not significantly increase the rate of floc formation (Figure 4.19). These results may indicate that, when using the PDA to optimize overall process performance, the optimum alum dose is the lowest dose beyond which FI does not increase more rapidly. In other words, if the value of the FI at one specific point were plotted vs. alum dose, the dose at which the slope of the trendline begins to approach zero would be the optimum. This approach is illustrated in Figure 4.20, in which FI at t = 180 sec. is plotted vs. each of the doses in Figures 4.18 and 4.19. The 180-sec. point was chosen for this particular set of coagulation conditions because it appeared to represent the relative kinetics of floc formation for each dose. A different point may be more appropriate for other water quality conditions and alum dose ranges. In Figure 4.21, PDA results are shown for optimized alum doses at each of the three pH levels used in the reservoir filtration experiments. The rapid floc formation kinetics observed at pH = 7.4 appear to correlate with the filtration data presented in Figure 4.15, which show that filtration performance can be optimized over a wider range of z.p. at higher pH conditions.

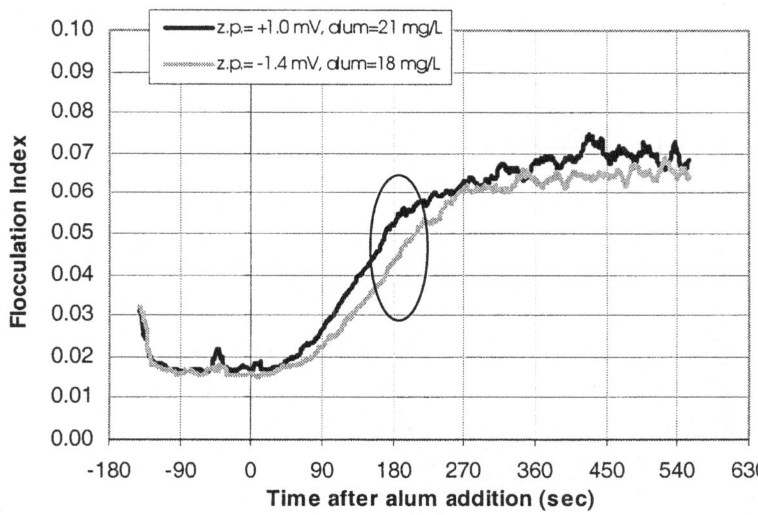

Figure 4.17 FI vs. time after alum addition in reservoir water (T = 8°C, pH = 6.15, TOC = 3.5 mg/L, turbidity = 15 ntu)

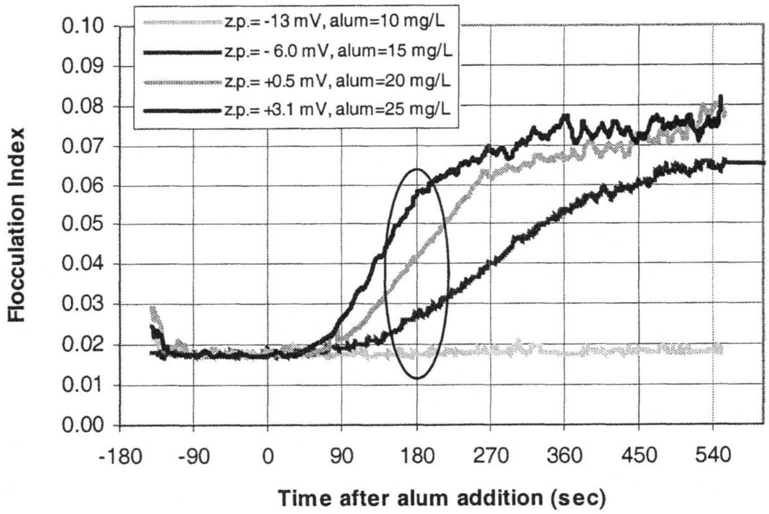

Figure 4.18 FI vs. time in reservoir water for optimum and non-optimum alum doses (T = 8°C, pH = 6.6 ± 0.1, TOC = 3.5 mg/L)

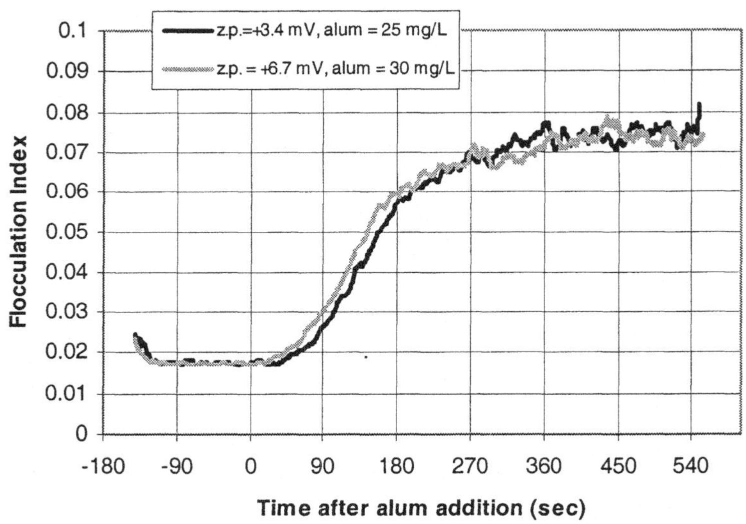

Figure 4.19 FI vs. time for optimum alum dose (25 mg/L) and overdose (30 mg/L) in reservoir water (pH = 6.6, TOC = 3.5 mg/L)

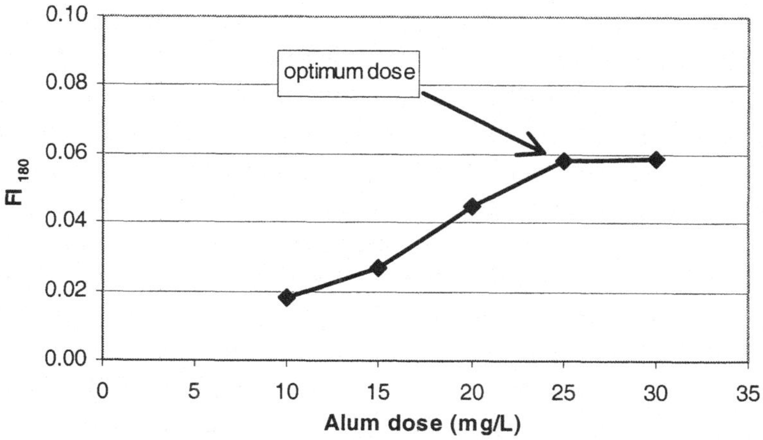

Figure 4.20 Alum dose vs. FI_{180} for data presented in Figures 4.16 through 4.18

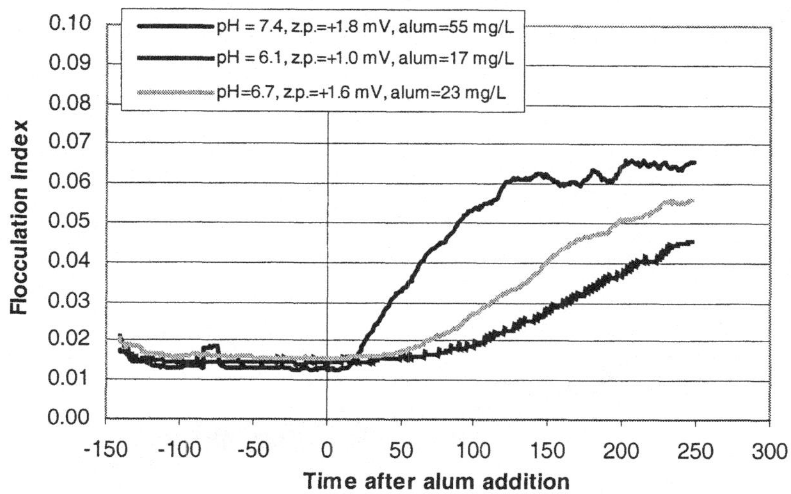

Figure 4.21 FI vs. time for three pH levels using optimized alum doses in reservoir water (T = 8°C)

TWO-STAGE ALUM ADDITION

Another objective of this project was to investigate two-stage coagulation processes, particularly for the treatment of high-NOM water. By use of a two-stage process, it was thought that the dual purposes of coagulation—removal of NOM and destabilization of suspended particles—could be accomplished more effectively. To optimize NOM removal, the target coagulation pH of the first stage was 6.0. To enhance particle removal, the pH was increased to 7.4 for the second stage. Although the research team was unable to apply this process during the runoff period, it was used during the reservoir water filtration experiments. The first set of two-stage runs (Figures 4.22 and 4.23) were conducted from October 23 through 25, 2001 (the single-stage data in Figure 4.15 were collected from October 29 through December 3, 2001). It should be noted that the water quality during this three-day span, although not equivalent to runoff-type water, presented its own challenges (Figure 3.3). Overall, the turbidity was significantly higher than normal (greater than 40 ntu) and subject to extreme fluctuations during a 24-hour period. Perhaps more importantly, the TOC concentration was also elevated during this period, by approximately 15 to 20 percent. These changes in the reservoir water quality forced the full-scale treatment processes to rely on blending with the river source, which during that time of year is characterized by very low-TOC, low-turbidity conditions. Figures 4.22 and 4.23, which present filtration results from experiments conducted in parallel, show that the two-stage process resulted in moderately improved filter performance when compared to the ambient process at pH = 6.7 (no acid/base addition). In Figure 4.23, steady-state particle counts are approximately 20 percent lower for the two-stage process. Towards the end of this phase of pilot experiments, the two-stage process was compared directly to the high-pH single-stage process (Figure 4.24). As anticipated, because of the high final pH of the two-stage process, settled turbidity values were approximately equal for the two-stage process (~1.2 ntu) and the high-pH single-stage process (Figure 4.15). Steady-state particle counts, however, were approximately 50 percent lower for the two-stage process.

Overall, the differences in steady-state particle counts between the two-stage and single-stage alum addition processes were not dramatic, as illustrated in Figures 4.22 through 4.24. However, use of the two-stage process did appear to improve filtration performance consistently.

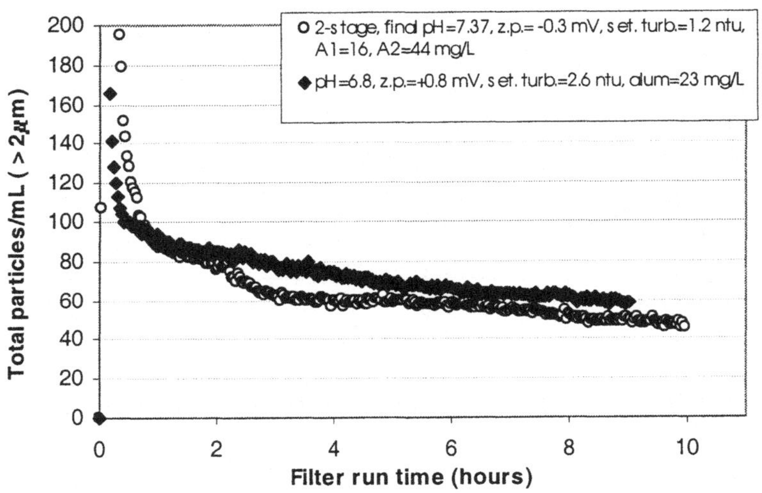

Figure 4.22 Comparison of filtration performance for two-stage and single-stage alum addition processes using reservoir water, October 23, 2001

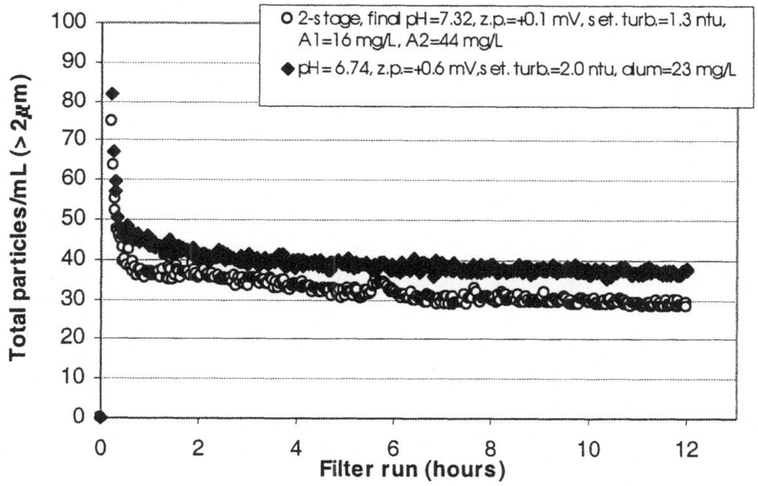

Figure 4.23 Comparison of filtration performance for two-stage and single-stage alum addition processes using reservoir water, October 24, 2001

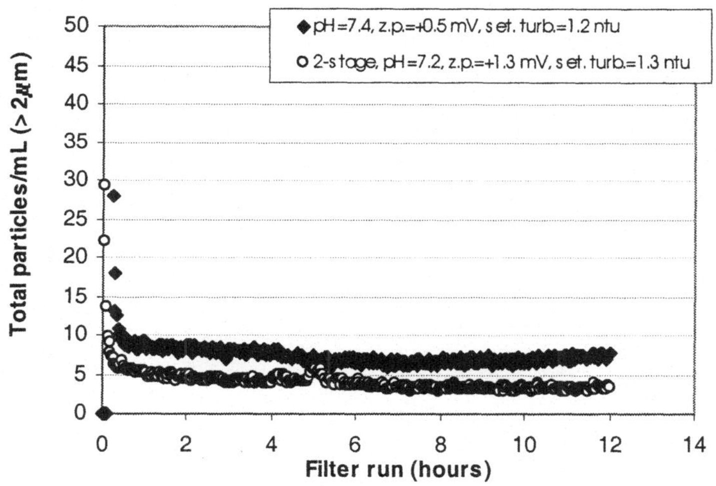

Figure 4.24 Comparison of filtration performance for two-stage and single-stage alum addition processes using reservoir water, December 3, 2001

CHAPTER 5
CONCLUSIONS

The primary objective of this research was to determine the impact of chemical sequencing and coagulation pH/zeta potential on filtration performance. The secondary goal was to determine effective coagulation processes—in terms of pH, zeta potential, and coagulant dose—for treating challenging (i.e., runoff-type) raw waters. Findings of the project included:

- In terms of overall process performance, the use of relatively high coagulation pH conditions along with alum doses optimized by zeta potential appeared to offer several advantages: 1) significantly improved particle removal in the settling process, 2) increased floc formation rates, 3) only slightly decreased TOC removal, compared to lower pH levels, and 4) a significantly wider operating range, in terms of zeta potential, for optimum filtration performance. The dual objectives of enhanced coagulation—optimization of NOM removal in addition to particles—can therefore be achieved at high pH as well as low pH.
- It does not appear, however, that filtration performance itself is significantly impacted by pH alone, as evidenced by the fact that excellent filter effluent quality was achievable from pH = 6.0 to 7.5 as long as alum doses were optimized by zeta potential. The primary effect of pH is on the alum dose required to achieve effective particle destabilization. At high pH and alum doses, the development of floc suitable for solid-liquid separation is clearly enhanced, perhaps due to a more consistent charge distribution on suspended particle surfaces. This mechanism would promote floc growth as well as increase the attachment efficiency of particles transported to filter media.
- Although filtration performance in general was improved at higher pH levels for both water sources, the effect was greater for the more challenging runoff water. Because higher pH significantly increases the optimum zeta potential range for filtration, a practical implication of this result is that use of a high coagulation pH process may be the most effective short-term treatment strategy for rapidly changing, runoff-type waters containing low alkalinity and high NOM levels. Clearly, a high-pH process would not be practical, in terms of chemical feed costs and solids handling issues, for raw water that is effectively treated by conventional processes. If a high-pH process is unfeasible for treating runoff water, the zeta potential of coagulated water should be continuously monitored and the coagulant dose adjusted accordingly.
- Settled water quality, although broadly predictive of filtration performance, was not always an accurate measure of overall process performance. In many cases, similar steady-state particle counts but significantly different settled turbidities were observed for high and low-pH processes. In other cases, filter effluent particle counts increased while settled turbidity decreased after changes to the coagulation conditions. These results reinforce the importance of proper particle conditioning, through effective coagulation processes, for overall process optimization.
- The PDA appears to be a useful alternative to jar tests for determining optimum coagulant doses and quickly evaluating different coagulation scenarios. Overall, processes that the PDA indicated were optimum, in terms of the kinetics of floc development, were shown to be optimum in terms of filtration performance as well.

- Two-stage coagulation processes may enhance filtration performance for a wide range of raw water quality conditions. In this study, two-stage processes moderately improved filter effluent particle counts in reservoir water experiments. In previous research, two-stage alum addition was shown to be significantly more effective for runoff water.

REFERENCES

Akitt, J.W., and A. Farthing. 1978. Studies of the Hydrolysis of the Aluminum (III) Cation. *Jour. Magnetic Resonance*, 32:345.

Amal, R., J.A. Raper, and T.D. Waite. 1992. Effect of Fulvic Acid Adsorption on the Aggregation Kinetics and Structure of Hematite Particles. *Jour. Coll. Interf. Sci.*, 151(1):244.

Amirtharajah, A., and K.M., Mills. 1982. Rapid-Mix Design for Mechanisms of Alum Coagulation. *Jour. AWWA*, 74:210.

Amirtharaja, A., and N. Tambo. 1991. Chapter 1. In *Mixing in Coagulation and Flocculation*, Denver, Colo.: AwwaRF.

AWWA Committee Report. 1989. Coagulation as an Integrated Water Treatment Process. *Jour. AWWA,* 81(10):72

Baes, C.F., and R.E. Mesmer. 1976. *The Hydrolysis of Cations*. New York, N.Y.: John Wiley and Sons.

Bell-Ajy, K., Abbaszadegan, M., Ibrahim, E., Verges, D. and LeChevallier, M. 2000. Conventional and Optimized Coagulation for NOM Removal. *Jour. AWWA*, 92(10):44.

Billica, J.A., and K. Gertig. 2000. Optimization of a Coagulation Process to Treat High TOC, Low Alkalinity Water and its Impact on Filtration Performance. In *Proc. 2000 AWWA WQTC*, Salt Lake City, UT.

Budd, G., A.D. Nickols, and G.A. Iversen. 1996. Optimization of Coagulation as a Multidimensional/Multiple Goal Treatment Process. In *Proc. of the AWWA Annual Conference*, Toronto.

Carlson, K.H., D. Gregory, and S. Macmillan. 1996. The Efficacy of Ozonation and Enhanced Coagulation for Treating Low Alkalinity, High Color Water. In *Proc. of the 1996 AWWA Annual Conference*, Toronto.

Carlson, K.H., and D. Gregory. 2000. Optimizing Water Treatment with Two-Stage Coagulation. *Jour. Environmental Engineering, ASCE,* 126(6):556.

Chipps, M.J., Logsdon, G.S., Hess, A.F., Bayley, R.G., Pressdee, J.R., and Rachwal, A.J. 2001. Practical Ways to Optimize Filter Performance. In *Proc. of Advances in Rapid Granular Filtration in Water Treatment* (International Conf., London). Denver, Colo.: AwwaRF.

Cleasby, J.L., Sindt, G.L., Watson, D.A., and Baumann, E.R. 1992. *Design and Operation Guidelines for Optimization of the High-Rate Filtration Process: Plant Demonstration Studies.* Denver, Colo.: AwwaRF.

Croue, J.P., Lefebvre, E., Martin, B., and Legube, B. 1993. Removal of Dissolved Hydrophobic and Hydrophilic Substances during Coagulation/Flocculation of Surface Waters. *Wat. Sci. Tech.*, 27(11):143.

Darby, J.L., and D.F. Lawler. 1990. Ripening in Depth Filtration: Effect of Particle Size on Removal and Head Loss. *Envir. Sci. Technol.*, 24(7):1069.

Dentel, S.K., and J.M. Gossett. 1988. Mechanisms of Coagulation with Aluminum Salts. *Jour. AWWA*, 80(4):187.

DiTommaso, J., and J.E. Van Benschoten. 1996. Path Dependence of Coagulation Reactions: An Experimental Study of Organic Matter Coagulation by Al and Fe Salts. In *Proc. of the AWWA Annual Conference*, Toronto.

Dittman, W., Grohmann, A., Klute, R., and Wiesmann, U. 1988. Mixing Systems for Pipe Reactors in Water Treatment. *Vom Wasser,* 70:129.

Edwards, M. 1997. Predicting DOC Removal During Enhanced Coagulation. *Jour. AWWA*, 89(5):78.

Edzwald, J.K., and J.E. Tobiason. 1999. Enhanced Coagulation: U.S. Requirements and a Broader View. *Wat. Sci. Tech.*, 40(9):63.

Edzwald, J.K., and J.E. Van Benschoten. 1997. Coagulation: Metal Hydrolysis, Solubility, and NOM Complexation. In *Proc. of the AWWA WQTC Conference*, Denver, Colo.

Elimelech, M., and C.R. O'Melia. 1990. Kinetics of Deposition of Colloidal Particles in Porous Media. *Envir. Sci. Tech.*, 24(10):1528.

Exall, K.N., and G.W. Vanloon. 2000. Using Coagulants to Remove Organic Matter. *Jour. AWWA*, 92(11):93.

Garrote, J.I., Bao, M., Castro, P., and Bao, M.J. 1995. Treatment of Tannery Effluents by a Two-Step Coagulation/Flocculation Process. *Wat. Res. Engr.*, 29(11):2605.

Gregory, J. 1984. Turbidity Fluctuations in Flowing Suspensions. *Jour. Coll. Interf. Sci.*, 105(2):357.

Habibian, M.T., and C.R. O'Melia. 1975. Particles, Polymers, and Performance in Filtration. *Jour. Env. Engr. Div., ASCE*, 101(EE4):567.

Hahn, H.H., and W. Stumm. 1968. Kinetics of Coagulation with Hydrolyzed Al (III): The Rate-Determining Step. *Jour. Coll. Interf. Sci.*, 28:134.

Hall, E.S. 1966. *Electrokinetic Study of Dispersions of Clay in Hydolyzed Aluminum Solutions.* Discussions Faraday Society, 42:197.

Huang C., and Chen, G. 1996. Use of the Fiber-optical Monitor in Evaluating the State of Flocculation. *Water Research*, 30(11):2723.

Hundt, T., and R. O'Melia. 1988. Aluminum-Fulvic Acid Interactions: Mechanisms and Applications. *Jour. AWWA*, 80(4):176.

Klute, R. 1990. Destabilization and Aggregation in Turbulent Pipe Flow. In *Chemical Water and Wastewater Treatment.* Edited by H.H. Kahn and R. Klute. Berlin: Springer-Verlag.

Lawler, D.F., and M. Benjamin. 2002. *Water Treatment.* Textbook. New York, N.Y.: McGraw-Hill.

Letterman, R.D., J.E. Quon, and R.S. 1973. Gemmell. Influence of Rapid-Mix Parameters on Flocculation. *Jour. AWWA*, 65:716.

Licskó, I. 1997. Realistic Coagulation Mechanisms in the Use of Aluminum and Iron (III) Salts. *Wat. Sci. Tech.*, 36(4):103.

Mackrle, S. 1962. Mechanisms of Coagulation in Water Treatment. *Jour. San. Engrg. Div.-ASCE*, SA3:1.

Matsui, Y., and N. Tambo. 1991. Online Floc Size Evaluation by Photometric Dispersion Analyzer. *Water Supply*, 9:71-78.

Matsui, Y., Yuasa, A., Furuya, Y., and Kamei, T. 1998. Dynamic Analysis of Coagulation with Alum and PACl. *Jour. AWWA*, 90(10):96.

Ohto, T., Zaitsu, Y., Matsui, Y., and Tambo, N. 1993. Advances in Photometric Dispersion Analyzer with Dual Wavelength Light. *Wat. Sci. Tech.*, 27(11):257.

O'Melia, C.R. 1985. Particles, Pretreatment, and Performance in Water Filtration. *Jour. Envir. Engr. Div., ASCE*, 111(6):874.

O'Melia, C.R., and W. Stumm. 1967. Theory of Water Filtration. In *Proc. of the AWWA Annual Conference*, Atlantic City, N.J.

Randtke, S.J. 1988. Organic Contaminant Removal by Coagulation and Related Processes. *Jour. AWWA*, 80(5):40.

Tambo, N., Matsui, Y., Kurotani, K., Kubota, M., Akiyama, H., Ohto, T., Zaitsu, Y., and Itoh, H. 1997. Control of Coagulation Processes by Dual Wavelength Particle Analyzer. *Wat. Sci. Tech.*, 36(4):135.

Tseng, T., B.D. Segal, and M. Edwards. 2000. Increasing Alkalinity to Reduce Turbidity. *Jour. AWWA*, 92(6):44.

Wahlroos, G. 1991. The Realization of Activated Carbon Filtration in a Water Purification Process for Highly Humic Water by Two-stage Iron(III) Salt Coagulation. *J. Water SRT-Aqua*, 40(4):211.

Wang, D., H. Tang, and J. Gregory. 2002. Relative Importance of Charge Neutralization and Precipitation on Coagulation of Kaolin with PACl: Effect of Sulfate Ion. *Envir. Sci. Technol.*, 36(8):1815.

Yao, K., M.T. Habibian, and C.R. O'Melia. 1971. Water and Waste Water Filtration: Concepts and Applications. *Env. Sci. Technol.*, 5(11):1105.

ABBREVIATIONS

α	filter attachment coefficient
A1	first alum dose in two-stage process
A2	second alum dose in two-stage process
AC	fluctuating component of voltage signal from PDA
Al	aluminum
alum	aluminum sulfate
AWWA	American Water Works Association
AwwaRF	Awwa Research Foundation
BOD	biochemical oxygen demand
°C	degrees Celsius
Co.	Company
COD	chemical oxygen demand
DC	average component of PDA voltage signal
DOC	dissolved organic carbon
DI	de-ionized
e.m.	electrophoretic mobility
FI	flocculation index
ft	feet
ft^2	square feet
G	velocity gradient
g/L	grams per liter
gpm	gallons per minute
gpm/ft	gallons per minute per square foot
HAA	haloacetic acids
III	iron
in.	inch
Jour.	Journal
L	liter
µeq/mg C	microequivalents per milligrams carbon
µm	micron
µm/s	micrometer per second
M	molar

mg/L	milligrams per liter
min.	minute
mL	milliliter
mm	millimeter
mV	millivolt
N	normality
NOM	natural organic matter
ntu	nephelometric turbidity units
PAC	project advisory committee
PDA	photometric dispersion analyzer
pH	logarithm of hydrogen ion concentratin
Proc.	Proceedings
PVC	Polyvinyl Chloride
rpm	revolutions per minute
sec	second
sols	sub-colloidal, poorly water-soluble, positively-charged aluminum hydroxide solids which form within 1 to 2 seconds following alum addition
t	time
T	temperature
THM	trihalomethanes
TOC	total organic carbon
TTHMFP	total tri-halomethane formation potential
UV_{254}	ultraviolet absorbance at 254 nm
V/cm	volts per centimeter
z.p.	zeta potential

IWA Publishing's authorised EU representative for General Product Safety Regulations is Diane D'Arras, 15 rue Duret, 75116 Paris, France, e-mail: safety@iwap.co.uk.

Printed and bound by CPI Group (UK) Ltd, Croydon, CR0 4YY
17/09/2026
02218813-0009